# Reviews of Environmental Contamination and Toxicology

VOLUME 241



More information about this series at http://www.springer.com/series/398

# Reviews of Environmental Contamination and Toxicology

Editor
Pim de Voogt



Founding Editor
Francis A. Gunther

Volume 241

Springer

ISSN 0179-5953 ISSN 2197-6554 (electronic)
Reviews of Environmental Contamination and Toxicology
ISBN 978-3-319-83624-9 ISBN 978-3-319-46945-4 (eBook)
DOI 10.1007/978-3-319-46945-4





Printed on acid-free paper

This Springer imprint is published by Springer Nature
The registered company is Springer International Publishing AG
The registered company address is: Gewerbestrasse 11, 6330 Cham, Switzerland

# Foreword

International concern in scientific, industrial, and governmental communities over traces of xenobiotics in foods and in both abiotic and biotic environments has justified the present triumvirate of specialized publications in this field: comprehensive reviews, rapidly published research papers and progress reports, and archival documentations These three international publications are integrated and scheduled to provide the coherency essential for nonduplicative and current progress in a field as dynamic and complex as environmental contamination and toxicology. This series is reserved exclusively for the diversified literature on "toxic" chemicals in our food, our feeds, our homes, recreational and working surroundings, our domestic animals, our wildlife, and ourselves. Tremendous efforts worldwide have been mobilized to evaluate the nature, presence, magnitude, fate, and toxicology of the chemicals loosed upon the Earth. Among the sequelae of this broad new emphasis is an undeniable need for an articulated set of authoritative publications, where one can find the latest important world literature produced by these emerging areas of science together with documentation of pertinent ancillary legislation.

Research directors and legislative or administrative advisers do not have the time to scan the escalating number of technical publications that may contain articles important to current responsibility. Rather, these individuals need the background provided by detailed reviews and the assurance that the latest information is made available to them, all with minimal literature searching. Similarly, the scientist assigned or attracted to a new problem is required to glean all literature pertinent to the task, to publish new developments or important new experimental details quickly, to inform others of findings that might alter their own efforts, and eventually to publish all his/her supporting data and conclusions for archival purposes.

In the fields of environmental contamination and toxicology, the sum of these concerns and responsibilities is decisively addressed by the uniform, encompassing, and timely publication format of the Springer triumvirate:

*Reviews of Environmental Contamination and Toxicology* [Vol. 1 through 97 (1962–1986) as Residue Reviews] for detailed review articles concerned with any aspects of chemical contaminants, including pesticides, in the total environment with toxicological considerations and consequences.

*Bulletin of Environmental Contamination and Toxicology* (Vol. 1 in 1966) for rapid publication of short reports of significant advances and discoveries in the fields of air, soil, water, and food contamination and pollution as well as methodology and other disciplines concerned with the introduction, presence, and effects of toxicants in the total environment.

*Archives of Environmental Contamination and Toxicology* (Vol. 1 in 1973) for important complete articles emphasizing and describing original experimental or theoretical research work pertaining to the scientific aspects of chemical contaminants in the environment.

The individual editors of these three publications comprise the joint Coordinating Board of Editors with referral within the board of manuscripts submitted to one publication but deemed by major emphasis or length more suitable for one of the others.

Coordinating Board of Editors

# Preface

The role of *Reviews* is to publish detailed scientific review articles on all aspects of environmental contamination and associated (eco)toxicological consequences. Such articles facilitate the often complex task of accessing and interpreting cogent scientific data within the confines of one or more closely related research fields.

In the 50+ years since *Reviews of Environmental Contamination and Toxicology* (formerly *Residue Reviews)* was first published, the number, scope, and complexity of environmental pollution incidents have grown unabated. During this entire period, the emphasis has been on publishing articles that address the presence and toxicity of environmental contaminants. New research is published each year on a myriad of environmental pollution issues facing people worldwide. This fact, and the routine discovery and reporting of emerging contaminants and new environmental contamination cases, creates an increasingly important function for *Reviews*. The staggering volume of scientific literature demands remedy by which data can be synthesized and made available to readers in an abridged form. *Reviews* addresses this need and provides detailed reviews worldwide to key scientists and science or policy administrators, whether employed by government, universities, nongovernmental organizations, or the private sector.

There is a panoply of environmental issues and concerns on which many scientists have focused their research in past years. The scope of this list is quite broad, encompassing environmental events globally that affect marine and terrestrial ecosystems; biotic and abiotic environments; impacts on plants, humans, and wildlife; and pollutants, both chemical and radioactive; as well as the ravages of environmental disease in virtually all environmental media (soil, water, air). New or enhanced safety and environmental concerns have emerged in the last decade to be added to incidents covered by the media, studied by scientists, and addressed by governmental and private institutions. Among these are events so striking that they are creating a paradigm shift. Two in particular are at the center of ever increasing media as well as scientific attention: bioterrorism and global warming. Unfortunately, these very worrisome issues are now superimposed on the already extensive list of ongoing environmental challenges.

The ultimate role of publishing scientific environmental research is to enhance understanding of the environment in ways that allow the public to be better informed or, in other words, to enable the public to have access to sufficient information. Because the public gets most of its information on science and technology from internet, TV news, and reports, the role for scientists as interpreters and brokers of scientific information to the public will grow rather than diminish. Environmentalism is an important global political force, resulting in the emergence of multinational consortia to control pollution and the evolution of the environmental ethic. Will the new politics of the twenty-first century involve a consortium of technologists and environmentalists, or a progressive confrontation? These matters are of genuine concern to governmental agencies and legislative bodies around the world.

For those who make the decisions about how our planet is managed, there is an ongoing need for continual surveillance and intelligent controls to avoid endangering the environment, public health, and wildlife. Ensuring safety-in-use of the many chemicals involved in our highly industrialized culture is a dynamic challenge, because the old, established materials are continually being displaced by newly developed molecules more acceptable to federal and state regulatory agencies, public health officials, and environmentalists. New legislation that will deal in an appropriate manner with this challenge is currently in the making or has been implemented recently, such as the REACH legislation in Europe. These regulations demand scientifically sound and documented dossiers on new chemicals.

*Reviews* publishes synoptic articles designed to treat the presence, fate, and, if possible, the safety of xenobiotics in any segment of the environment. These reviews can be either general or specific, but properly lie in the domains of analytical chemistry and its methodology, biochemistry, human and animal medicine, legislation, pharmacology, physiology, (eco)toxicology, and regulation. Certain affairs in food technology concerned specifically with pesticide and other food-additive problems may also be appropriate.

Because manuscripts are published in the order in which they are received in final form, it may seem that some important aspects have been neglected at times. However, these apparent omissions are recognized, and pertinent manuscripts are likely in preparation or planned. The field is so very large and the interests in it are so varied that the editor and the editorial board earnestly solicit authors and suggestions of underrepresented topics to make this international book series yet more useful and worthwhile.

Justification for the preparation of any review for this book series is that it deals with some aspect of the many real problems arising from the presence of anthropogenic chemicals in our surroundings. Thus, manuscripts may encompass case studies from any country. Additionally, chemical contamination in any manner of air, water, soil, or plant or animal life is within these objectives and their scope.

Manuscripts are often contributed by invitation. However, nominations for new topics or topics in areas that are rapidly advancing are welcome. Preliminary communication with the Editor-in-Chief is recommended before volunteered review manuscripts are submitted. *Reviews* is registered in WebofScience™.

Inclusion in the Science Citation Index serves to encourage scientists in academia to contribute to the series. The impact factor in recent years has increased from 2.5 in 2009 to almost 4 in 2013. The Editor-in-Chief and the Editorial Board strive for a further increase of the journal impact factor by actively inviting authors to submit manuscripts.

Amsterdam, The Netherlands
January 2015

Pim de Voogt

# Contents

# Contributors

**H. Allmendinger** Currenta GmbH and Co. OHG, Leverkusen, Germany

**Paula M.C. Antunes** AquaTox Testing & Consulting Inc., Guelph, ON, Canada

**B.T.A. Bossuyt** Huntsman Europe, Everberg, Belgium

**S. Chatterjee** Defence Research Laboratory, DRDO, Tezpur, Assam, India

**S. Datta** Defence Research Laboratory, DRDO, Tezpur, Assam, India

**Nancy D. Denslow** Department of Physiological Sciences and Center for Environmental and Human Toxicology, University of Florida, Gainesville, FL, USA

**Camille Dumat** Centre d'Etude et de Recherche Travail Organisation Pouvoir (CERTOP), UMR5044, Université J. Jaurès—Toulouse II, Toulouse Cedex 9, France

**D.K. Gupta** Gottfried Wilhelm Leibniz Universität Hannover, Institut für Radioökologie und Strahlenschutz (IRS), Hannover, Germany

**B. Hidding** BASF SE, Ludwigshafen, Germany

**Sana Khalid** Department of Environmental Sciences, COMSATS Institute of Information Technology, Vehari, Pakistan

**Girish Kumar** Faculty of Fisheries and Protection of Waters, South Bohemian Research Center of Aquaculture and Biodiversity of Hydrocenoses, University of South Bohemia in Ceske Budejovice, Vodňany, Czech Republic

**Nabeel Khan Niazi** Institute of Soil and Environmental Sciences, University of Agriculture Faisalabad, Faisalabad, Pakistan

**T. Schupp** Faculty of Chemical Engineering, Muenster University of Applied Science, Steinfurt, Germany

**Muhammad Shahid** Department of Environmental Sciences, COMSATS Institute of Information Technology, Vehari, Pakistan

**B. Tury** Global Isocyanates Limited, Manchester, UK

**A.V. Voronina** Department of Radiochemistry and Applied Ecology, Physical Technology Institute, Ural Federal University, Ekaterinburg, Russia

**C. Walther** Gottfried Wilhelm Leibniz Universität Hannover, Institut für Radioökologie und Strahlenschutz (IRS), Hannover, Germany

**R.J. West** The Dow Chemical Company, Midland, MI, USA

# Gene Expression Profiling in Fish Toxicology: A Review

**Girish Kumar and Nancy D. Denslow**

## Contents



G. Kumar (✉)
Faculty of Fisheries and Protection of Waters, South Bohemian Research Center of Aquaculture and Biodiversity of Hydrocenoses, University of South Bohemia in Ceske Budejovice, Zátiší 728/II, 389 25 Vodňany, Czech Republic
e-mail: girishkumar.nio@gmail.com

N.D. Denslow
Department of Physiological Sciences and Center for Environmental and Human Toxicology, University of Florida, Gainesville, FL 32611, USA
e-mail: ndenslow@ufl.edu

 
P. de Voogt (ed.), *Reviews of Environmental Contamination and Toxicology, Volume 241*, Reviews of Environmental Contamination and Toxicology 241,
DOI 10.1007/398_2016_10

# 1 Introduction

Pollution of the aquatic environment, due to the growing levels of contaminants associated with human activity such as industry, mining, agriculture, and household waste production, has become a serious problem worldwide (Conti et al. 2012; Copat et al. 2012a, b). Aquatic organisms, such as fish, accumulate pollutants directly from contaminated water and indirectly via the food chain (Sasaki et al. 1997). Bio-concentration of potentially harmful substances—e.g. heavy metals, pesticides, polychlorinated biphenyls (PCBs), polycyclic aromatic hydrocarbons (PAHs) etc., in fish present a major threat to human health. Since fish occupy the top of the aquatic food chain, they are widely used as a bio-indicator to evaluate the health of aquatic ecosystem (Camargo and Martinez 2007).

Use of fish in toxicity studies can be traced back to 1800s when Penny and Adams (1863) used the chemicals (used in dye-works) to study their effects on goldfish (*Carassius auratus*). Although basic research in toxicology began in the 1800s, test methodology was established, especially on fish, only in the 1940s (Hunn 1989). Subsequently, considerable effort was made and a standardized test method was published in 1960 (American Public Health Association et al. 1960). Since the 1960s, rapid development in toxicity testing changed toxicology from a descriptive science to one in which mechanisms of action of toxicants are recognized. In subsequent decades, with increased emphasis on the use of molecular biology techniques, toxicological studies further expanded in many areas including chemical carcinogenesis and xenobiotic metabolism, among many others. In recent years, analysis of gene expression has become the method of choice for many toxicological studies. This is due to the fact that toxicant exposure often alter gene expression profile as cells regulate certain genes to protect cellular structures and repair damage (Causton et al. 2001). Analysis of these gene expression data can provide information regarding contaminant exposure and potential effects at higher biological levels (Garcia-Reyero et al. 2008). Gene expression profile, therefore, provides a sensitive measurable endpoint for toxicity and thus can serve as early warning of a specific biological endpoint.

Until 1992, Northern blot and subtractive hybridization were the only methods used in gene expression analysis. Although these methods are still quite useful and reliable, they are tedious, time consuming and difficult to perform. Additionally, they require large amounts of mRNA that can be limited in many situations. In 1992, polymerase chain reaction (PCR) based alternative methods—real time quantitative PCR (RT-qPCR) (Porcher et al. 1992) and differential display reverse transcription PCR (DDRT-PCR) (Liang and Pardee 1992) were developed. These methods allow the identification of differentially expressed genes in a less labor-intensive fashion than the methods mentioned above. RT-qPCR now represent the method of choice for analyzing gene expression of a small number of genes in thousands of samples. In the last decade, use of microarrays and next-generation sequencing (NGS) technologies paved the way for a paradigm shift from the analysis of a selected set of genes to a whole-genome screening of potentially all expressed genes. However, so-called close-end transcriptomic technique, microarray require prior knowledge of gene sequences and are therefore unsuitable for non-model species where limited or no sequence information is available. To overcome this limitation, pyrosequencing technology was utilized to create a transcriptome-wide microarray from a non-model species (Garcia-Reyero et al. 2008). Nevertheless, in recent years, NGS technology has become a method of choice for transcriptome profiling due to its decreasing costs and high-throughput.

In past, many reviews were published to discuss and describe the various aspects of fish toxicology using a particular model species (Ankley and Villeneuve 2006), technique (Ju et al. 2007; Williams et al. 2014) or toxicant (Larkin et al. 2003; Garcia-Reyero and Denslow 2006). However, no attempt has been made to present a comprehensive review on application of gene expression in fish toxicology. Therefore, the objective of this paper is to review the application of gene expression profiling techniques in fish toxicology and to discuss the biological insights already gained from these techniques in fish species.

## 2 Use of Northern Blotting in Fish Toxicology

Northern blotting was one of the first molecular biology techniques used in fish toxicology to determine the relative abundance of gene transcripts in response to toxicants (Chen 1983). In this technique, electrophoretically separated bands of RNA are hybridized with DNA fragments labeled with radioactive nucleotides complementary to the gene(s) of interest to determine the differentially expressed gene (Alwine et al. 1977).

In the beginning, Northern blot analysis was used to study the transcriptional induction of vitellogenin (VTG) gene in response to endocrine disrupting chemical (EDC), 17β-estradiol ($E_2$) (Chen 1983). Since then, use of Northern blots have been expanded to a large number of EDCs to study the gene expression profile of many biomarker genes (Flouriot et al. 1997; Arukwe et al. 2001, 2002; Bowman et al. 2002; Yadetie and Male 2002; Larkin et al. 2003).

Vitellogenin, the egg yolk precursor protein, is synthesized in all oviparous vertebrates. It is normally produced by females in response to estradiol during oogenesis. However, it also gets synthesized in livers of males upon exposure to estrogen or an estrogen mimic chemical. This characteristic of VTG has been explored using several fish species for detecting the presence of xenoestrogens from contaminated aquatic environments (Sumpter and Jobling 1995; Folmar et al. 1996; Denslow et al. 1997a, b). In order to find out the presence of estrogenic compounds in effluents of pulp and paper mills, juvenile whitefish (*Coregonus lavaretus* L. *s.l.*) were exposed to diluted effluents from three operating pulp, paper, and paperboard mills in Southern Lake Saimaa, Finland (Mellanen et al. 1999). Northern blot analysis showed increased VTG mRNA levels suggesting that the effluent of pulp and paper mills were a source of estrogenic contaminants. In addition to the VTG mRNA, the presence of estrogen or estrogen mimic chemicals in aquatic environments also cause increase in expression levels of choriogenin mRNA (Murata et al. 1997; Arukwe et al. 2001) and estrogen receptor (ER) mRNA (Flouriot et al. 1997; Bowman et al. 2002). Therefore, these genes can be used as a potential biomarker to identify the presence of estrogenic chemicals in aquatic environment.

Alkylphenols such as 4-nonylphenol (NP) are one of the wide variety of environmental chemicals that have estrogenic effects. The estrogenic effects of NP have been assessed through gene expression levels of VTG and zona radiata (ZR) proteins (Jobling et al. 1996; Arukwe et al. 1997; Oppen-Berntsen et al. 1999; Jung et al. 2006). However, to enhance the understanding of the mechanisms of actions and to develop new potential biomarkers for xenoestrogens, Yadetie and Male (2002) studied the effects of NP (50 and 125 mg/kg body weight) on mRNA levels of leutinizing hormone beta (LHβ) in male and female juvenile of Atlantic salmon (*Salmo salar*). Northern blot analysis revealed significant induction of mRNA levels of LHβ in female juvenile salmonids (up to sixfold), suggesting that NP has the potential to perturb the regulation of LHβ gene expression by mimicking $E_2$ and thus can be used as a biomarker for pituitary effects of xenoestrogens. Nevertheless, one should always take into consideration that too much circulating $E_2$ can signal the pituitary to shut down by feedback inhibition (Kriegsfeld et al. 2006).

Northern blot analysis has also been used to study the down-regulation of biomarker genes in response to exposure to chemicals. For instance, treatment of mature female tilapia (*Oreochromis niloticus*) with high levels of androgen (17-α-methyltestosterone) (20 mg/kg, in 50 % ethanol/saline (5 mg/ml)) resulted in a pronounced decline in VTG mRNA levels and serum $E_2$ level (Lazier et al. 1996). These results suggest the presence of an inhibitory site of action at the hypothalamic-pituitary axis, probably through an androgen receptor or through an estrogen receptor after local aromatization of 17α-methyltestosterone (Lazier et al. 1996). In another study, authors observed reduction in transthyretin (TTR) expression in males of gilthead seabream (*Sparus aurata*) after estrogen treatment (Funkenstein et al. 2000). Furthermore, Northern blot analysis of hepatic retinol-binding protein (RBP) revealed reduced level of mRNA expression in adult male gilthead seabream (Funkenstein 2001) and rainbow trout (Sammar et al. 2001) after $E_2$ treatment. Since the $E_2$ levels in fish during the period between vitellogenic growth and final oocyte maturation decreases, the down-regulation of RBP and TTR genes in response to estrogen treatment in the above studies indicate the role of TTR and RBP in transport of maternal thyroid hormones and retinol (Funkenstein 2000, 2001).

Cytochrome P450s play a pivotal role in metabolism of toxic compounds in vertebrates and hence the expression of P450s influenced by exposure to xenoestrogens and other environmental contaminants (Williams et al. 1998). The effect of $E_2$ and testosterone on expression of P450s (CYP2M1, CYP2K1, CYP3A27) was studied in rainbow trout (Buhler et al. 2000). Gene expression analysis suggested that exposure of fish to endocrine disruptors can adversely affect a number of physiological processes through mechanisms involving altered levels of expression of specific P450 isozymes mRNA (Buhler et al. 2000). Furthermore, a higher levels of cytochrome P450 1A1 (CYP1A) mRNA was observed when fish were exposed to β-naphthoflavone (Leaver and George 2000) and polybrominated diphenyl ethers (Boon et al. 2002). The reason for increased levels of CYP1A transcripts in above mentioned studies is due to the fact that polycyclic aromatic hydrocarbons (PAHs) transcriptionally activate CYP1A by binding to Aryl hydrocarbon Receptor (AhR). In another study, Ferraris et al. (2005) attempted to investigate the molecular mechanisms of the dicarboximide fungicide iprodione (Ip) in rainbow trout. Findings of the study revealed that Ip-mediated CYP1A3 gene induction involves the activation of AhR complex via phosphorylation-dephosphorylation reactions.

Northern blot analysis is not only used to measure expression level of mRNA but also to validate gene induction profiles of DDRT-PCR and microarrays. One of the main advantages of this method is that it can detect the size of the mRNA and alternatively spliced transcripts (if present). However, this technique can be used to measure the expression levels of only a small number of genes. In addition, it is labor intensive and have limited resolution. Therefore in recent years, it has been largely replaced by newer methods, such as DDRT-PCR, SSH, RT-qPCR, microarrays and NGS.

# 3 Use of DDRT-PCR in Fish Toxicology

The technique of DD RT-PCR was developed by Liang and Pardee in 1992. Since then this method has gained popularity among researchers and is now used to identify and isolate differentially expressed genes in several experimental systems. In this technique reverse transcription of mRNA is done with anchored oligo-dT primers, followed by PCR amplification using anchor primer and a limited number of short arbitrary primers. Amplified products are then visualized either by radioactivity or fluorescence.

In fish toxicology, DDRT-PCR has been applied to examine the alterations in gene expression following chemical exposure in natural and experimental environments. For example, gene expression patterns in sheepshead minnows (*Cyprinodon variegatus*) were determined thorough *in vivo* exposure to natural and anthropogenic estrogens (Denslow et al. 2001a, b). Based on the results authors concluded that $E_2$, $EE_2$ (17α-ethinyl estradiol), and DES (diethylstilbestrol) act through the same pathway when administered through the water at low concentrations. In largemouth bass changes in gene expression induced by exposure to natural estrogen (Bowman et al. 2002) and paper and pulp mill effluents (Denslow et al. 2004) were measured. Differential expression of genes (VTG, ER, ERp72 ZP2, fibrinogen γ) in response to $E_2$ at different time series (0, 6, 12, 24, 48 h) confirmed the idea that gene regulation by $E_2$ is complicated and time dependent (Bowman et al. 2002). Measurements of changes in gene expression induced by exposure to paper mill effluents resulted in an up-regulation of CYP1A and down-regulation of genes normally expressed during the reproductive season in females. These antiestrogenic changes suggest that effluents of paper and pulp mills could decrease the reproductive success in affected populations (Denslow et al. 2004).

Carginale et al. (2002) investigated the ability of cadmium (Cd) to affect gene transcription in Antarctic icefish (*Chionodraco hamatus*) using DDRT-PCR. Results indicated that Cd has both inhibitory effect as well as positive modulator of gene expression. The inhibitory effect of Cd may be due to an improper folding of zinc finger domains and/or oxidation of the cysteine residues while positive effect may be attributed to the fact that binding of Cd to zinc containing proteins release this metal which act as a positive modulator of gene expression. DDRT-PCR has also been used to study the contaminant effects of PAHs such as anthracene (Peterson and Bain 2004) and pyrene (Roling et al. 2004) in mummichogs (*Fundulus heteroclitus*). Genes demonstrating differential expression after PAHs (pyrene and anthracene) exposure suggested that multiple biomarkers are required to get insight into the mechanism of toxicity and information on potential effects of toxicants. In another study, altered gene expression in killifish (*F. heteroclitus*) was studied by collecting the samples from a highly contaminated site (Elizabeth River, VA, USA), and non-contaminated reference site (King's Creek, VA, USA). Seventy-four differentially expressed mRNAs were identified including sex and population specific suggesting that sex of the fish and population differentiation play a role in response to contaminant exposure (Meyer et al. 2005).

In tilapia, lindane exposure caused differential expression of 50 different genes, four of these showed homology with immunoglobulin heavy chain, coagulation factor V, casein kinase 2 alpha, and receptor protein- tyrosine-like phosphatase (Colli-Dula et al. 2009). These findings suggest that lindane triggers the expression of genes that are involved in immunological and stress responses in fish. Woźny et al. (2012) used DD PCR to study the molecular background of zearalenone (ZEA) toxicity in rainbow trout. A total of 59 differentially expressed cDNA were identified, five of these genes (β-centractin, bccip, α-enolase, protein C, and ferritin heavy chain subunit) were successfully confirmed by RT-qPCR. Differential expression of above mentioned genes indicates that ZEA interferes with blood coagulation and different cellular components (i.e. cytoskeleton).

The DD RT-PCR is an effective and powerful approach for detecting novel genes involved in toxicant effect pathways and identifying new molecular biomarkers. However, it has not been widely applied in fish toxicology research due to its several inherent limitations. One of the major limitations is the large percentage of false-positive results generated by the PCR step. In addition, the identification of differentially expressed genes remains tedious and confirmation of the gene expression is required by RT-qPCR. For these reasons, in the mid-2000s, DDRT-PCR technique was superseded by RT-qPCR, DNA microarrays and RNA-Seq.

# 4 Use of SSH in Fish Toxicology

Suppression subtractive hybridization (SSH) is a method used to distinguish between two mRNA samples that are either overexpressed or exclusively expressed in one sample compared to another. This technique is based on selective amplification of target cDNA and simultaneously suppressing the amplification of non-target cDNA (Diatchenko et al. 1996). SSH combines normalization and subtraction in a single procedure. This dramatically increases the probability of obtaining low-abundance differentially expressed cDNA. However, one of the major limitations of SSH is that it compares gene transcripts from only two samples at one time and therefore it is time consuming and labor intensive. Another limitation is that the process is leaky and thus increases the probability of obtaining non-specific genes. Therefore, everything needs to be validated by RT-qPCR. Nevertheless, SSH has been employed by many research groups to investigate various aspects of aquatic toxicology. For example, Brown et al. (2004) studied the response of male plaice (*Pleuronectes platessa*) to $EE_2$ exposure. The results showed a selective down-regulation of some egg proteins (VTG, ZRP) mRNA responsible to maintain the quality of eggs. Thus, exposure of fish to $EE_2$ may cause reproductive failure by diminishing the quality of eggs. In the subsequent year, Volz et al. (2005) assessed the qualitative gene expression changes in whole brain, liver, and testis of adult male Japanese medaka in response to 2, 3, 7, 8-Tetrachlorodibenzo-*p*-dioxin (TCDD). CYP1A mRNA was found significantly higher in TCDD-exposed brain and liver. However, testis specific CYP1A mRNA was not significantly induced, suggesting that TCDD has tissue specific effect on fish. In another study, Li et al. (2008) examined the effects of exposure to polychlorinated biphenyls (PCBs) on false kelpfish (*Sebastiscus marmoratus*). Functional analysis of differentially expressed genes were found to be associated with long-term potentiation and neurotransmitter release, neuroendocrine, mitosis and cell proliferation, energy-related metabolism, general metabolism, signal protein, hemopoiesis system and immune system. However, only 45 of the 108 gene sequences could be identified with homologous database sequences. Therefore, further investigation is required to understand the effects of PCB-induced toxicity in fish. Furthermore, SSH was applied to isolate and enrich the differentially expressed genes from juvenile samples of European seabass (*Dicentrarchus labrax*) collected at the reference site and at a contaminated site. Altered expression in genes involved in energy metabolism, immune system activity and antioxidant response was identified (Nogueira et al. 2010). Based on the results, the authors suggested that these changes due to environmental contamination could affect the population in future.

In order to study the toxic effects of Cd, Dangre et al. (2010) identified over 700 Cd responsive cDNAs in sheepshead minnow. These genes were found to be broadly involved in protective and repair functions as well as cellular and metabolic processes. In addition, the authors found that Cd exposure down-regulated both embryonic α and β globin in sheepshead minnow. These results support the perspective that Cd exerts its toxicity through the modulation of multiple physiological processes.

To elucidate the genes in response to tetrodotoxin (TTX), Matsumoto et al. (2011) examined the hepatic gene expression profile of marine puffer fish (*Takifugu rubripes*) by intramuscular administration of 0.50 mg TTX/kg. A high level of up-regulation was observed in hepcidin precursor gene. In addition, induction of complement C3, serotransferrin, apolipoprotein A-1, high temperature adaptation protein Wap65-2, complement C7, fibrinogen beta chain, and heat-shock protein 4 were also obtained. Increased expression of these genes confirmed that the exposure of TTX increases the gene expression of the acute-phase response proteins. SSH was also used to identify alterations in gene transcription of silver carp (*Hypophthalmichthys molitrix*) and bighead carp (*Aristichthys nobilis*) after exposure to Microcystin-LR (MCLR). In silver carp, sequencing analysis and homology searches revealed 75 unique differentially expressed genes in response to MCLR (Qu et al. 2011). Of the 75 unique genes, 38 were identified with functions of immune response, transport and cell metabolism. Some genes (*Fs59*, *Fs70*, *Rs15*, and *Rs2*) showed the time-dependent transcriptional in silver carp livers. In the case of bighead carp, Cai et al. (2012) tested the gene expression profile of cytidine deaminase (*cda*) gene. Altered expression profile of the *cda* gene was inversely proportional to time and directly proportional to dose of MCLR in most of the tested tissue. These findings indicate that the effects of MCLR are highly complex and in future more cellular pathways affected by MCLR could be identified.

## 5 Use of RT-qPCR in Fish Toxicology

Reverse transcription-quantitative PCR (RT-qPCR) is an accurate and sensitive tool to analyze and validate differential gene expression. This technique came into light in early 1990s (Higuchi et al. 1993) and became popular among researchers for analyzing gene expression of a moderate number of genes in anywhere from a small number to thousands of samples. Despite the limitation in the number of genes that can be analyzed per sample, RT-qPCR has been widely used to study the toxic effects of a large number of chemical pollutants including heavy metals, natural and synthetic steroids, pesticides, herbicides, fungicides, PCBs, perfluorononanoic acid (PFNA), PAHs, dioxins, microcytin, brominated and organophosphorus flame retardants, pharmaceuticals and personal care products, and manufactured nanoparticles.

### *5.1 Endocrine Disrupting Chemicals (EDCs)*

To determine the effects of EDCs on gene expression, Celius et al. (2000) quantified the mRNA levels in $E_2$ treated (0.01, 0.1, 1.0 or 10 mg/kg body weight) rainbow trout and reported induction of ZR and VTG mRNA. However, induction of ZR-gene occurred at all concentrations of $E_2$ while VTG was only significantly induced at the highest dose (10 mg/kg). The more sensitive nature of ZR suggests that ZR-gene is more suitable indicator than VTG to estrogenic substances. Hogan et al. (2008) investigated the sensitivity of three hormone responsive genes (VTG, spiggin, ARβ) in threespine stickleback (*Gasterosteus aculeatus*) by exposing the fish to 1, 10, and 100 ng/L of methyltestosterone or $E_2$ for 7 days. Tissue-specific induction in expression of VTG (liver) and spiggin (kidney) mRNA in male and female fish respectively suggest that expression pattern of these genes can be used to identify compounds causing reproductive dysfunction in fishes.

Furthermore, effects of EDCs (bisphenol A, nonylphenol, octylphenol) and trace metals (Cd, Cu, and Zn) were investigated in different tissues (brain, eye, gonad, intestine, liver, muscle, and skin), development stages (embryonic and larval), and gender types (hermaphrodites and secondary males) of mangrove killifish (*Kryptolebias marmoratus*) (Rhee et al. 2009). The RT-qPCR analysis showed suppressive effects of EDCs on expression of metallothionein (MT) gene. Based on the findings, authors suggested that the expression level of MT may be affected if metals are present in water along with EDCs. In another study, exposure of sex steroids (estradiol, 2-methyl-testosterone) in Chinese rare minnow (*Gobiocypris rarus*) showed up-regulation of *foxl2* gene after treatment with estradiol and down-regulation after treatment with 2-methyl-testosterone (Jiang et al. 2011). These results suggest that *foxl2* mRNA expression is regulated by sex hormones, and therefore *foxl2* can be used as a molecular indicator in monitoring for environmental endocrine disruptors.

In an attempt to study the effects of $E_2$ (10, 100, and 1000 μg/L) on stress related genes, Woo et al. (2012) investigated the transcriptional changes in intestinal, liver and muscle tissues of Japanese medaka. RT-qPCR analysis showed a tissue specific alteration in mRNA levels of catalase (CAT), glutathione peroxidase and ubiquitin genes. The differential expression of a range of stress-related genes in this study suggests that transcriptional changes in these genes could be used as a rapid assay to study the effects of $E_2$ exposure (Woo et al. 2012).

Attempts were also made to study the effects of EDCs on embryonic development and sexual differentiation. For instance, Lei et al. (2013) used RT-qPCR to evaluate reproductive toxicities in Japanese medaka by exposing the fertilized eggs to 1, 10, 100, and 1000 ng/L of β-estradiol 17-valerate (EV) for 15 days. Results showed an up-regulation of VTG-I in liver of males while down-regulation of ERα and VTG-I in liver of female individuals. These findings suggest that EV is a reproductive toxicant and cause reproductive failure in fish species.

In another study, zebrafish (*Danio rerio*) embryos and larvae were used to assess how EDCs ($EE_2$, genistein, fadrozole) interfere with embryogenesis (Santos et al. 2014). RT-qPCR analysis revealed alterations in the expression levels of the hormone receptors genes (*esr* and *ar*) and apoptosis-related genes (*p53* and *c-jun*) at different time points (between 2 and 144 h post-fertilization), suggesting a bidirectional communication between hormones action and apoptosis pathways on early development. These results also suggest that exposure to EDCs poses a threat to aquatic organisms during embryogenesis by impairing physiological processes essential to a normal development and, consequently, survival.

In a more recent study, Schiller et al. (2014) proposed a novel method of determining the effects of substances with estrogenic and anti-androgenic activity by comparing the embryos of two species (medaka and zebrafish). Patterns of gene expression between the two species were compared by exposing the embryos (medaka: 7-day post fertilization; zebrafish: 48 and 96 h post fertilization) to estrogenic ($EE_2$) and anti-androgenic (flutamide) reference compounds and five test compounds with endocrine activities (bisphenol A, genistein, prochloraz, linuron, and propanil). Estrogenic responses were comparable between the two species while responses to anti-androgenic chemicals differed between medaka and zebrafish. While responses to anti-androgenic chemicals were more sensitive in medaka embryos, a broader range of morphological effects were observed in zebrafish embryos. Therefore, together medaka and zebrafish embryos would be beneficial for testing effects of EDCs in early life states.

## 5.2 *Metals, Metallic Nanoparticles and Organometals*

To assess the risk of metal ions in fishes inhabiting metal contaminated area, MT gene has been used as a biomarker. For example, metal ions ($Cu^{2+}$, $Cd^{2+}$, $Hg^{2+}$, $Ni^{2+}$, $Pb^{2+}$, $Zn^{2+}$) exposure in *Tilapia* showed increased hepatic and gills MT mRNA levels. However, renal MT mRNA levels did not show significant induction in response to $Cd^{2+}$ and $Zn^{2+}$ metal ions, suggesting that gene expression of MT is tissue dependent (Cheung et al. 2004). In another study, Chan et al. (2006) investigated the induction of MT gene in zebrafish following the administration of different metal ions ($Cu^{2+}$, $Cd^{2+}$, $Hg^{2+}$, and $Zn^{2+}$). RT-qPCR analysis showed highest level of MT mRNA induction (40–50 fold) in embryo-larvae in response to $Hg^{2+}$ followed by $Cd^{2+}$. However, in the *in vitro* study of liver cell-line (ZFL), $Cd^{2+}$ was the most potent inducer of MT mRNA (up to 250 folds), followed by $Cu^{2+}$ and $Zn^{2+}$ (50–100 folds). These results showed that ZFL is more sensitive to metal exposure and thus it is a better endpoint to study metal contamination. Furthermore, MT gene expression was studied in metal exposed (Cu, Zn, Cd) mangrove killifish. Gene expression analysis showed that Cd is strongest inducer of MT gene (Rhee et al. 2009). The contrasting results obtained in above mentioned studies suggest that induction of MT gene in response to metals is influenced by tissue, sex, and species of fish.

In order to understand the effect and mode of action (MoA) of uranium (U), juvenile Atlantic salmon were exposed 0.25, 0.5 and 1.0 mg/L waterborne depleted uranyl acetate (Song et al. 2012). The hepatic transcriptional response showed induction of genes involved several major toxicological pathways including the anti-oxidant defense, DNA damage and repair, apoptosis and protein degradation. Although no significant adverse effects were observed after relatively short exposure (48 h), results suggested that long term exposure to U may cause adverse effects such as DNA damage in aquatic organisms.

The transcriptional response to mercury chloride ($HgCl_2$) exposure in zebrafish larvae revealed that a concentration between 20 and 30 μg/L in aquatic environment may cause oxidative stress by modulating activity and expression of P2X7 receptor (Cruz et al. 2013). Cu is known to induce apoptotic cell death in fish species. To find out the apoptotic pathways triggered by Cu, zebrafish were exposed to 12.5 and 100 μg /L of Cu during 6, 12, 24 and 48 h. Based on the mRNA expression analysis, authors proposed that Cu initiate apoptosis via intrinsic pathway (caspase-9), through p53 activation; then by the extrinsic pathway (caspase-8) and finally by the caspase-independent pathway (AIF) (Luzio et al. 2013). In another study, Jing et al. (2013) evaluated the potential and applicability of heat shock protein (HSP70) gene to monitor heavy metal stress (Cu, Cd) in white cloud mountain minnow (*Tanichthys albonubes*). The results indicated dose-dependent expression of HSP70 mRNA. However, gene expressions did not show consistent changes between transcription and translation levels. Therefore, both transcriptional and translational variation mechanisms should be taken into account while analyzing metal stress response pathways. In a more recent study, Arini et al. (2015) demonstrated that genetic response to Cd and Zn contamination is tissue- and metal-dependent. Based on the results of the study, authors suggested that in case of binary contamination (Zn and Cd), Zn plays a protective role and prevents against oxidative stress while Cd induces oxidative stress and generate detoxification mechanisms.

To investigate the toxic effects of manufactured nanomaterials (MNs) in aquatic environment, Pham et al. (2012) studied gene expression pattern of eight stress related genes (*MT, GST, p53, HSP70, TF, Orla C3-1, Chg-L,* and *vtg-I*) after exposing Medaka to silver nitrate or a silver nanoparticle (Ag-NP). Significant induction of MT and GST (Glutathione S-transferase) mRNA suggest that the Ag-NPs cause oxidative and inflammatory stress in fish. Similarly, increase in expression of CAT, GST, and SOD (Superoxide dismutase) mRNA were observed when juvenile tilapia were exposed to titanium-dioxide NPs ($TiO_2$-NPs) (Varela-Valencia et al. 2014). However, exposure to rutile phase $TiO_2$-NPs generated a faster and stronger response in expression of CAT and GST gene than anatase. This suggests that rutile causes the production of higher levels of ROS than anatase.

## 5.3 *Pesticides and Herbicides*

The intensive use of pesticides and herbicides in modern agricultural system has resulted in serious environmental problems. Researchers have demonstrated that these agricultural chemicals adversely affect the immune system, early developmental stages and sexual differentiation in fish species (Li et al. 2013a; Zhang et al. 2013; Glisic et al. 2016; Wang et al. 2015a). Gene expression analysis using RT-qPCR has shown that exposure of fish to chlorpyrifos (CPF) and atrazine (ATR) causes down-regulation of acetylcholinesterase (AChE) mRNA, an enzyme that helps in regulation of cholinergic nervous transmission (Xing et al. 2010). Thus, the alteration in expression of AChE mRNA could be used to reveal toxic mechanisms related to cholinergic signaling. In another study, CPF exposure to common carp resulted in increased IgM and complement C3 expression, indicating that CPF causes immunotoxicity to common carp (Li et al. 2013a). Transcriptional profiles of zebrafish embryos (24 h post-fertilization) after an acute exposure (48 and 72 h) to ATR resulted in significant changes in *cyp1a, cyp3a65* genes (Glisic et al. 2016). However, ATR failed to perturb transcription of GST isoforms and antioxidant enzymes (AOE) genes. This suggests that GST and AOE genes are not reliable markers of acute ATR exposure.

Contamination of water by organochlorine pesticides and brominated and organophosphorus flame retardants also causes toxicity in fish species. For example, induction of VTG mRNA levels in response to these chemicals suggests that they have estrogenic activity and can impair fish reproduction (Chow et al. 2011; Wang et al. 2015a). Monocrotophos (MCP), an organophosphorus pesticide has also been found to have significant estrogenic properties. MCP exposure in zebrafish stimulated *foxl2* and suppressed *dmrt1* transcription factors (Zhang et al. 2013). Differential expression of these genes indirectly elevated the gonadal (*cyp19a1a*) and brain (*cyp19a1b*) aromatase gene expression. Gonadal aromatase promotes the phenotypic feminization while brain aromatase, which is thought to be involved in reproductive physiology via the brain-pituitary-gonadal axis, causes indirect impact on sexual differentiation. MCP exposure also induces $E_2$ levels and promotes steroidogenesis by inducing transcription of CYP11A1, CYP17, and CYP19A genes (Wang et al. 2015b). Flame retardants (brominated and organophosphorus) have been identified to cause cardiac toxicity in fish (Wu et al. 2013; Du et al. 2015). For example, hexabromocyclododecane (HBCD) exposure of zebrafish embryos resulted in elevated expression of the *Tbx5* and *Nkx2.5* genes, which are mainly responsible for alteration of heart rate (Wu et al. 2013). Bisphenol-A (BPA), a derivative of bisphenol was found to be toxic in early development of fish. The up-regulation of *sp4* in response to BPA suggests its role in altering brain development (Lam et al. 2011).

## 5.4 *Miscellaneous*

Dioxins such as 2, 3, 7, 8-tetrachlorodibenzo-*p*-dioxins (TCDDs) are of great concern as environmental contaminants. TCDD cause a wide range of adverse effects in vertebrates. In general, fishes are highly sensitive to TCDD exposure, especially during early developmental stages (Peterson et al. 1993). RT-qPCR analysis in seabream (*Pagrus major*) embryos revealed that TCDD-induced early stage toxicity is caused by increased expression of *rsAHR2* and *CYP1A* genes (Yamauchi et al. 2006). Additionally, TCDD exposure causes modest changes in expression of microRNAs (*miR-451, 23a, 23b, 24 and 27e*) that are critical for hematopoiesis and cardiovascular development (Jenny et al. 2012). Furthermore, TCDD causes cardiac toxicity in fish embryos by suppressing transcription of a battery of cell cycle related genes (Chen 2015). RT-qPCR analysis has also been used in determining the toxic effects of PCBs in fishes by measuring the CYP1A mRNA transcript abundance (Dixon et al. 2002; Zanette et al. 2009).

Gene expression profiling in response to pharmaceuticals and their metabolites and compounds used in personal care products have shown that these contaminants cause adverse effects on aquatic organisms through the cellular toxicity, *p53* related genotoxicity and estrogenic effects (Hong et al. 2007; Yamauchi et al. 2008). In aquatic environment fish are also exposed to toxins such as microcystin (MC) and cylindrospermopsin (CYN) released by cyanobacterial blooms. Gene expression analysis have shown that MC disrupts thyroid hormone homeostasis in zebrafish (Yan et al. 2012). In another study, Li et al. (2013b) suggested that metabolization and detoxification of MCs in zebrafish is done by increased expression of CYP3A65 and its receptor PXR (pregnane X receptor) gene. Authors also found that expression of CYP3A65 and PXR gene is regulated by miRNA (dre-miR-27b). CYN exposure has been shown to alter the activity and gene expression of some of the glutathione related enzymes in tilapias (Gutiérrez-Praena et al. 2013). The elevated mRNA levels of GST in liver and kidney of tilapia indicates that CYN cause hepatocyte and renal injury to fish.

Fan et al. (2010) characterized ten nervous system genes (*syn2a, mbp, gap43, elavl3, nkx2.2a, ngn1, α1-tubulin, nestin, shha, gfap*) to study the effects of ethanol in developmental neurotoxicity. The expression profiles of these genes were altered after exposure to sublethal doses of ethanol, a known developmental neurotoxicant, to zebrafish embryos/larvae. Presence of some of the affected genes (*α1-tubulin, ngn1, mbp,* and *gfap*) exclusively in the nervous system suggest that gene expression profile to these genes can be used as an endpoint for efficient screening for developmental neurotoxicity using zebrafish. Yuan et al. (2014) evaluated the toxic effects of benzo[a]pyrene (BaP) on developmental stage of Chinese rare minnows. Simultaneous up-regulation of the *cyp* (*cyp1a, 1b1*, and *1c1*) and ATP-binding cassette (ABC) transporter (*abcc1, abcc2*, and *abcg2*) gene transcripts suggests that these genes work in a coordinated manner to detoxify the effects of BaP in early developmental stages of fish.

To study the oxidative stress response of nitrite, gene expression profile of antioxidant genes were evaluated in Wuchang bream (*Megalobrama amblycephala*). The acute nitrite exposure in juvenile fish resulted in the formation of excess of reactive oxygen species (ROS) and reduced expression of antioxidant genes (*MaCAT, MaGPx1, MaCu/Zn-SOD*). The reverse correlation between ROS and antioxidant genes suggest that imbalance between ROS and antioxidant defenses is the mechanism underlying nitrite toxicity in Wuchang bream (Sun et al. 2014). In another study, toxic effect of perfluorononanoic acid (PFNA) was examined in zebrafish larvae. Gene expression analysis showed an up-regulation of genes involved in oxidative stress and apoptotic pathways (Yang et al. 2014). Furthermore, to evaluate the adverse effects and intrinsic toxicological properties of tributyltin (TBT), Tian et al. (2015) studied the gene expression profile of aromatase gene in male guppies (*Poecilia reticulata*). RT-qPCR analysis revealed that TBT exposure suppresses the expression of two isoforms of aromatase. Suppression of aromatase causes the disturbance in reproductive behavior of fish by endocrine-disrupting action.

## 6 Use of Microarrays in Fish Toxicology

Microarrays (also commonly known as DNA arrays, DNA chip or biochip) are a high throughput technology that are used to measure expression levels of thousands or even tens of thousands of genes simultaneously. This allows researchers to study the effects of toxicants on the expression of virtually all known genes within a single microarray (Ramsay 1998). However, one of the major limitations for the use of microarrays in fish toxicology is the lack of sequence information for non-model species. Nevertheless, ESTs and custom arrays based on genes expressed in response to specific contaminants have been developed and used with microarray to provide the genome wide expression data. Although use of microarray technology in molecular biology research started in mid-1990s (Ramsay 1998), the first use of microarray in fish was done by Gracey and co-workers in 2001 (Gracey et al. 2001). Since then microarrays have been used in large number of fish species to evaluate the gene expression responses to toxicants and to identify genes and molecular pathways involved in fish responses to toxicants.

### *6.1 Metals, Metallic Nanoparticles and Organometals*

In European flounder (*Platichthys flesus*) microarrays were applied to investigate the effect of acute exposure to Cd (Sheader et al. 2006). The mRNA expression analysis detected up-regulation of 27 transcripts (including Cu/Zn SOD, thioredoxin, a peroxiredoxin, and a GST) and down-regulation of CYP1A. Alteration in these genes indicates that Cd exposure cause oxidative stress in flounder. In another study, fibroblast cell line (SAF1) from gilthead bream was exposed to sub-toxic levels of Cu, Zn, and Cd (Minghetti et al. 2011). Microarrays analysis showed an increased mRNA levels of MT and GST genes in response to Cu, Zn, and Cd. However, only ATP7A mRNA levels were increased in Cu treated SAF1. Findings of this study suggest that expression of copper transporting ATPase, ATP7A, may be regulated at the transcriptional level directly by Cu.

The toxic effect and mode of action (MoA) of uranium (U) was studied in the brain of the zebrafish by exposing the fish to 15 and 100 μg/L of waterborne U for 3 and 10 days (Lerebours et al. 2010). A total of 56 transcripts expressed differentially and were categorized into eight functional classes with the most significantly affected being the olfactory system. Results of this study suggest that genes in the olfactory region may be used as a sensitive transcriptional biomarkers of U waterborne exposure. Song et al. (2014) expanded their previous study (Song et al. 2012) of U toxicity in Atlantic salmon using microarrays in combination with RT-qPCR. Hepatic transcriptional responses showed that U exposure affects various toxicity pathways involving mitochondrial function, oxidative stress, nuclear receptor signaling, and organ damage. In another study, genome-wide gene expression responses to metal poisoning and mechanistic insights into metal toxicity were evaluated in whole adult male zebrafish following exposure to three metals (nickel, cobalt, chromium) (Baer et al. 2014). Based on the expression data authors identified changes in gene expression of group of genes involved in adaptive responses, cell cycle regulation, apoptosis and metabolic depression.

Research has shown that aggregated nanoparticles and release of dissolved metal ions produce different toxicity effects in aquatic organisms. For example, global gene expression analysis in gills of zebrafish following exposure to nanocopper, nanosilver, and nanotitania produced a distinct gene expression profile for each of nanometal or soluble metal, suggesting that each exposure is producing biological response by a different mechanism (Griffitt et al. 2009). Jovanovic et al. (2011) examined the acute toxicity of engineered nanoparticles (titanium dioxide/$TiO_2$ and hydroxylated fullerenes/$C_{60}$ $(OH)_{24}$) in zebrafish embryos. The nanoparticle exposure caused changes in genes related to circadian rhythm, cell kinase activity, intracellular trafficking, and immune response. However, expression patterns were similar for down-regulated genes but different for upregulated genes. Finding of this study suggest that different nanomaterials may have similar suppression but different up-regulation patterns of transcriptome changes. Furthermore, sublethal effects of nano-Ag and dissolved Ag were examined on rainbow trout (Gagné et al. 2012). Gene expression analysis revealed that genes involved in inflammation are sensitive to nano-Ag exposure while oxidative stress and protein stability related genes are sensitive to dissolved Ag. This shows that the toxicity of Ag differ depending on the presence of Ag nanoparticles and aggregates.

Methylmercury (MeHg), a known neurotoxicant, is widely distributed contaminant polluting many aquatic environments. To study its toxic effects, fathead minnows (*Pimephales promelas*) were fed with food containing MeHg (Klaper et al. 2006). Gene expression analysis revealed a marked up-regulation in VTG mRNA in males and a significant decline in VTG gene expression in female fish. This differential expression indicates that MeHg cause endocrine disruption in fish. In an expansion to this study, Klaper et al. (2008) used 15,000 gene microarray to investigate the impact of acute and chronic MeHg exposures in male gonad and liver tissues of fathead minnows. Gene expression analysis showed significant differences in differentially expressed genes in each of the tissues in response to acute and chronic treatment. In general, genes associated with stress, adaptation, and the immune system were altered. In a more comprehensive study, Yadetie et al. (2013) prepared Atlantic cod 135 k oligonucleotide arrays to study the mechanisms of toxicity and range of molecular targets of MeHg. In addition to the known effects of MeHg such as oxidative stress, gene transcripts coding for enzymes involved in metabolism of amino acids, fatty acids and glucose were up-regulated. These results suggest that exposure to MeHg may cause disruption of nutrient metabolism in fish species.

## 6.2 Endocrine Disrupting Chemicals

Usefulness of microarrays has been also demonstrated in studying the effects of EDCs in fish species. For example, Larkin et al. (2003) used estrogen-responsive gene array of sheepshead minnow (SHM) to study the effects of $E_2$, $EE_2$, DES, *p*-nonylphenol (PNP), methoxychlor (MXC), and endosulfan (ES). Gene expression analysis showed that $E_2$, $EE_2$, DES, and MXC have similar genetic signatures. However, the number of genes examined in this study were limited (30 genes). Knoebl et al. (2006) expanded the existing gene array for SHM and used it to examine temporal changes in gene expression for male SHM exposed to 100 ng $E_2$/L for five time points between 0 and 48 h. Authors identified five patterns of temporal induction in genes involved in oocyte development. However, this difference in temporal expression was lost at higher concentration (500 ng $E_2$/L) probably due to feedback effects in brain. Findings of this study suggest that exposure to high levels of environmental contaminants may affect the normal ordered expression of genes required for reproduction.

In another study, Moens et al. (2007a) developed a cDNA microarray of 398 different gene fragments for the evaluation of endocrine disruption in common carp (*Cyprinus carpio*). Gene expression profiles of fish treated with $E_2$ resulted in a total of 35 differentially expressed genes, whereas cortisol exposure affected only three genes. These results indicate the discriminating power of the developed array and its usefulness to describe the toxicological mode of action of EDCs. In order to identify novel genomic responses to $EE_2$ exposure, Martyniuk et al. (2007) utilized microarrays in conjunction with RT-qPCR. Gene expression analysis in liver and telencephalon of male zebrafish identified tissue specific gene responses. For example, genes involved in protein metabolism, carbohydrate metabolism were down-regulated in the liver but were induced in the telencephalon.

Furthermore, a cDNA microarray consisting of 9692 clones was developed for threespined stickleback and used to assess gene expression responses to the dibenzanthracene (DbA) and $E_2$ as single and binary treatments (Geoghegan et al. 2008). Induction of CYP1A and ERα mRNA was observed in both male and female individuals in response to DbA and $E_2$ respectively. However, exposure to $E_2$ and binary mixture showed a different expression pattern in male and female sticklebacks for VTG, zona radiata and chorion protein mRNA. VTG, zona radiata and chorion protein mRNA was induced in male fish exposed to $E_2$ and a binary mixture while females showed induction only with the binary mixture. The disparity in expression pattern of genes in male and female in response to $E_2$ is most likely because of high endogenous levels of $E_2$ in female fish (Geoghegan et al. 2008).

Wei et al. (2008) used a custom cDNA microarray containing 1773 unique genes to study the effects of perfluorooctanoic acid (PFOA) in male and female rare minnows. Gene expression analysis revealed the inhibition of thyroid hormone biosynthesis genes and induction of estrogen-responsive genes strongly suggesting that PFOA is an EDC.

Wang et al. (2011) investigated the effects of trilostane on the transcriptional regulatory dynamics of zebrafish. Differential expression analysis revealed that trilostane-induced genes are broadly involved in cell proliferation, differentiation, migration, and apoptosis. Additionally, authors concluded that trilostane-induced effects on fish endocrine functions are not confined to the HPG-axis alone. Further, to strengthen the toxicogenomic study, Pham et al. (2011) constructed cDNA microarrays from Japanese medaka to screen biomarkers for $E_2$, NP, and 2-chlorophenol (2CP). To study the toxicological potency of DES in fish, Adedeji et al. (2012) exposed sexually mature fathead minnows of both sexes to 1, 10, or 100 ng of DES/L of water in a flow-through system. Evaluation of hepatic transcript profile revealed induced mRNA level of VTG and ERα in both sexes at DES concentrations $\geq$10 ng/L suggesting concentration dependent change in gene expression.

## 6.3 *Miscellaneous*

Exposure of fish to TCDD not only causes alteration in gene expression but multiple injuries in the heart, liver, kidney, intestine, and ovary (Volz et al. 2006; Liu et al. 2014). Microarray analysis of 175-gene array from medaka revealed that alteration in hepatic gene expression and histological changes are strongly dependent on dose and time (Volz et al. 2006). Liu et al. (2014) showed that altered genes in response to TCDD were involved in pathways associated with cardiac necrosis/cell death, cardiac fibrosis, renal necrosis/cell death, and liver necrosis/cell death. In another study, Hofsteen et al. (2013) revealed that TCDD exposure inhibit the regeneration of heart muscle and also alters the pattern of gene expression induced by wounding at the ventricle in zebrafish. Based on the results authors speculated that TCDD exposure inhibits a process that is important for progenitor cells needed for both early heart development and regeneration.

Utility of microarray in longjaw mudsucker (*Gillichthys mirabilis*) revealed that exposure of fish to hypoxic environment first suppresses the energy-requiring processes like protein synthesis and locomotion in skeletal muscle and then induces genes in liver needed for anaerobic ATP production and for gluconeogenesis (Gracey et al. 2001). Martinovic et al. (2009) showed that hypoxia caused alteration in reproduction of zebrafish by reorganization of lipid transport and mechanisms involved in hydroxyl steroid dehydrogenase alterations and down-regulation of contractile elements.

In order to assess the potential of microarray technique for environmental monitoring, Williams et al. (2003) sampled European flounder from the polluted

Tyne estuary. The inductions of CYP1A, UDP-glucuronosyltransferase (UDPGT) and aldehyde dehydrogenase mRNA in the liver of flounder suggested that Tyne estuary is contaminated with polycyclic aromatic hydrocarbons (PAHs). Moens et al. (2007b) conducted a chronic toxicity test by exposing common carp to effluent for 21 days under flow-through conditions. Microarray analysis revealed that effluent treatment affected molecular pathways associated with the energy balance of the fish, including changes in carbohydrate and lipid metabolism, as well as digestive enzyme activity. Furthermore, Lie et al. (2009) investigated the effects of contaminants on gene expression in two natural populations of Atlantic cod from western Norway. The up-regulation of CYP1A, heme oxygenase, ferritin, and metallothionein gene suggests the heavy metal pollution in sampling sites.

Other studies using microarray technology in fish include the examination of the toxic effects on development and survival of zebrafish embryos in response to a wider range of ecotoxicants (2, 4-dichlorophenol, 3, 4-dichloroaniline, pentachlorophenol, and cadmium chloride) (Sawle et al. 2010), dithiocarbamates (DTCs) (van Boxtel et al. 2010), flusilazole (Hermsen et al. 2012), dimethylformamide (DMF), dimethylsulfoxide (DMSO) (Turner et al. 2012) and PAHs (Goodale et al. 2013; Jayasundara et al. 2015). Tilton et al. (2005) used rainbow trout as a research model for study of a fungicide metabolite, aflatoxin B1 (AFB1) induced hepatocarcinogenesis. The altered gene expressions were found to be similar to those observed in humans and other mammalian models. Thus, rainbow trout can be used as model to study the carcinogenesis in humans and other mammals. Martyniuk et al. (2010) examined the toxic effects of sub-chronic dietary exposures to 2.95 mg dieldrin/kg feed in the neuroendocrine brain of largemouth bass. Microarrays, proteomics, and pathway analysis revealed that sub-chronic exposure to dieldrin alters the abundance of mRNAs and proteins in the hypothalamus that are associated with cell metabolism, cell stability and integrity, stress, and DNA repair. The application of microarrays has also been used for the characterization of gene expression patterns associated with exposure to ethynylestradiol, 2,2,4,4-tetrabromodiphenyl ether, diquat, chromium VI, BaP (Hook et al. 2006), 2, 4-dinitrotoluene (2, 4-DNT) (Wintz et al. 2006), perfluorooctane sulfonate (PFOS) (Hagenaars et al. 2008), domoic acid (Lefebvre et al. 2009), polyfluorinated and perfluorinated compounds (PFCs) (Wei et al. 2009), cyclotrimethylenetrinitramine (RDX) (Gust et al. 2011) and brominated flame retardant mixture (Williams et al. 2013).

## 7 Use of NGS in Fish Toxicology

The advent of next-generation sequencing (NGS) technologies has driven a massive acceleration in research and development. The decreasing per base cost of raw sequence and increasing throughput has made it affordable even for non-model species. Although microarrays are still the primary technology for transcriptome analysis, in recent years RNA-seq using NGS has been widely used to study the effects of chemicals and pollutants on fish species. This is largely because RNA-seq does not require prior sequence knowledge and is able to detect even low expressing genes. This increased sensitivity allows for the identification of novel biomarkers. In addition, RNA-seq experiments require less RNA input compared to microarrays (Wang et al. 2009).

In one of the first toxicological studies in fish using NGS, Pierron et al. (2011) examined the effects of Cd and Cu exposure on yellow perch (*Perca flavescens*). Transcriptome analysis in treated fish revealed a decrease in mRNA levels of genes associated with protein biosynthesis, immune system, and lipid and energy metabolism. Based on the results the authors suggested that down-regulation of these genes could be due to an impairment of bile acid metabolism by Cd and silencing of genes by epigenetic modification of histones and DNA. In a subsequent year, Huang et al. (2012) studied the transcriptional response of marine medaka (*Oryzias melastigma*) embryos after PFOS exposure. Differential expression of genes related to neurobehavioral defects, mitochondrial dysfunction, and the metabolism of proteins and fats provided the useful data for identifying affected pathways and possible mechanisms of PFOS in marine fish. In another study, the toxicological effects of TCDD, a well-known teratogen, on development of zebrafish was examined using RNA-seq supplemented with microarrays and RT-qPCR (Jenny et al. 2012). Expression profiles of miRNA suggested that TCDD exposure causes abnormal phenotypes associated with hematopoiesis and cardiovascular development. In a similar study, RNA-seq was used to identify miRNAs as potential biomarker for monitoring stress responses towards acidic aluminum-rich water in Atlantic salmon (Kure et al. 2013).

Garcia de la serrana et al. (2012) successfully applied the RNA-seq technology to study the effects of deepwater horizon (DH) oil release on gulf killifish (*Fundulus grandis*). Altered expression of a large set of genes (1070 up and 1251 down-regulated) indicates that exposure to DH oil mediates a broad and complex genomic response including activation of AHR and related pathways associated with xenobiotic hydrocarbons. RNA-seq has also been used to investigate sublethal effects of oil sands process-affected water (OSPW) in fathead minnows (Wiseman et al. 2013). Based on abundances of transcripts in livers of fathead minnows exposed to untreated OSPW, authors proposed that OSPW cause toxicity through a mechanism involving reactive oxygen species (ROS) and apoptosis.

In order to find out the transcriptional responses to Hg, liver transcriptome of gadoid fish (*Brosme brosme*) was sequenced (Olsvik et al. 2013). Findings of this study suggested that Hg exposure affect the lipid metabolism and beta-oxidation in liver. Webster et al. (2013) used the RNA-seq to investigate the molecular mechanisms of metal (water-borne mixture of metals, including copper and zinc) tolerance in brown trout (*Salmo trutta*). Based on functional analysis, authors suggested that metal- and ion-homeostasis pathways could be the most important mechanisms contributing to the metal tolerance in brown trout. Du et al. (2014) extended the RNA-seq technology to study the gene expression in response to loading stress in Japanese grenadier anchovy (*Coilia nasus*). Based on pathway enrichment analysis of loading stress-responsive unigenes, authors concluded that loading stress induces liver injury through the mitochondrial apoptosis pathway. In a most recent study, Bahamonde et al. (2015) used RNA-seq in conjunction with microarray to study the molecular signaling cascades associated with intersex in rainbow darter (*Etheostoma caeruleum*) exposed to municipal wastewater. Differential expression analysis of genes involved in sex differentiation (*sox9, foxl2,* and *dmrt1*) and reproduction (*esr1, esrb, ar,vtg, cyp19a1,* and *cyp11a*) showed an intermediate expression level in intersex males when compared to phenotypic males and females. Thus, RNA-seq can be used to discriminate pollutant-exposed males without intersex from those males with intersex.

## 8 Toxicogenomics Databases and Bioinformatics

A rapid progress in toxicogenomics approaches in the last decade has generated a wealth of gene expression data. There was a growing desire in scientific community to make these data sets publicly available once research findings have been published. Efforts were made in this direction and several public databases (ArrayExpress, Gene Expression Omnibus (GEO), Comparative Toxicogenomics Database (CTD), Center for Information Biology gene Expression database (CIBEX), Sequence Read Archive (SRA)) for functional genomics were established. The ArrayExpress (http://www.ebi.ac.uk/arrayexpress) (Brazma et al. 2003) and GEO (http://www.ncbi.nlm.nih.gov/geo/) (Edgar et al. 2002) are repositories for microarray and high-throughput sequencing based genomics experiments. The CTD (http://ctdbase.org/) (Mattingly et al. 2003) was developed to formalize, harmonize and centralize the information on numerous genes and proteins responding to environmental toxic agents across diverse species.

To facilitate and promote the sharing of high quality and well-annotated microarray data within the scientific community, members of Microarray Gene Expression Data (MGED) Society (http://www.mged.org) laid down and published standardizations for microarray experiments and data description called MIAME (Minimum Information About a Microarray Experiment). In Japan, along with the MGED activity, Ikeo et al. (2003) developed a gene expression database called CIBEX, with a data retrieval system in compliance with MIAME. In 2009, database for next-generation sequencing data, SRA, was established as a part of the International Nucleotide Sequence Database Collaboration (INSDC) (Shumway et al. 2010). Similar to MIAME guidelines for microarray data, guidelines for high-throughput sequencing data, Minimum Information about Sequencing Experiment (MINSEQE; http://www.fged.org/projects/minseqe/) were proposed.

Some other toxicogenomics resources such as External RNA Controls Consortium (ERCC) was developed to verify technical performance and to interpret quality of gene expression data. To further strengthen the toxicogenomics resources, MicroArray Quality Control project (MAQC) was initiated which is working to improve application of the microarray and next-generation sequencing technologies by providing large reference datasets and developing guidelines for toxicogenomics data analysis. In another project called Percellome Toxicogenomics Project (PTR), a systematic method to normalize the mRNA expression values from DNA microarrays and RT-qPCR in a "per one cell" basis was developed. This method helps to monitor the time- and dose-dependent alteration of gene expression in response to various chemicals.

In toxicogenomics, bioinformatics approaches are used to identify chemical classes, toxic effects of a new compounds and pathways affected by a chemical stressors. The clustering method is used to compare gene expression profile of two samples or relative intensity of a single sample in response to a particular chemical. This approach groups the genes within or between samples to find the chemicals with similar toxic mechanisms (Waring et al. 2001). The connectivity mapping approach is used to find the toxic effects of a new chemical by comparing the gene expression profile to the compounds with similar gene expression profile whose toxicity is already known. Bioinformatics tools are also used to discern mechanisms of action of toxic responses. In mechanistic analysis, gene annotation databases such as Gene Ontology (GO, http://geneontology.org/page/go-database), pathway databases such as Kyoto Encyclopedia of Genes and Genomes (KEGG, http://www.genome.jp/kegg/) and WikiPathways (http://www.wikipathways.org) are commonly used. However, in recent years CTD has evolved as premier toxicology resource for mechanistic analysis of toxicogenomics data (Davis et al. 2015). Although the quantity of toxicogenomics data are increasing at an exponential rate, there is a lack of capacity in computational and bioinformatics tools. Therefore, new and efficient data-mining programs and comparative genomic tools are required to explore and interpret data in a time frame compatible with the rate at which they are generated.

## 9 Summary and Conclusion

Gene expression analysis is one of the most widely used techniques in detecting and elucidating molecular responses to toxicants. Over the last 10 years by the introduction and widespread adoption of toxicogenomic techniques in fish toxicology has made a paradigm shift from studying expression patterns of single or few molecular endpoints to genome wide expression response. Nevertheless, toxic effects of chemicals and pollutants have been investigated on only a few fish species (Tables 1, 2, 3, 4, 5, and 6). This is largely because of limited genomic information available for non-model fish species. Even for fish species in which the genome has been sequenced, poor annotation of the genome often poses difficulty in interpretation of the biological significance of differentially expressed gene sets. In recent years, however, great efforts have been made on sequencing the genomes of economically important aquaculture species. In this effort 12 fish species have been fully sequenced and assembled (http://www.ensembl.org). In addition, 22 fish species have been annotated by the NCBI Eukaryotic Genome Annotation Pipeline (http://www.ncbi.nlm.nih.gov). Furthermore, recent advances in NGS technologies

**Table 1** Examples of studies using Northern blotting in fish toxicology

| Toxicants/compounds | Species | Tissue | Target gene | Major findings | Reference |
|---|---|---|---|---|---|
| 17α-methyltestosterone (17αMT) | *Oreochromis niloticus* | Liver | Vitellogenin (VTG) | Inhibition of VTG gene expression | Lazier et al. (1996) |
| Effluents from pulp, paper, and paperboard mills | *Coregonus lavaretus* | Liver | VTG | Increased vitellogenin gene expression | Mellanen et al. (1999) |
| 17β-estradiol ($E_2$) | *Sparus aurata* | Liver | VTG, transthyretin (TTR) | Induction of VTG mRNA level and reduction of TTR mRNA levels by $E_2$ treatment | Funkenstein et al. (2000) |
| β-naphthoflavone (BNF) | *Pleuronectes platessa* | Gill, heart, liver | Cytochrome P450 1A (CYP1A) | Up-regulation of CYP1A after BNF treatment, with greatest response in gill followed by heart and liver | Leaver and George (2000) |
| $E_2$, nonylphenol (NP) | *Oncorhynchus mykiss* | Liver | Zona radiate (ZR), VTG | Up-regulation of ZR and VTG mRNA | Arukwe et al. (2002) |
| $E_2$, ethinylestradiol ($EE_2$) | *Cyprinodon variegatus* | Liver | VTG | Induction of VTG mRNA in liver | Bowman et al. (2000) |
| $E_2$ | *Micropterus salmoides* | Liver | VTG, estrogen receptor (ER) | Hepatic ER and VTG mRNA are coordinately regulated following acute $E_2$ exposure (2 mg/kg) | Bowman et al. (2002) |
| Polybrominated diphenyl ethers (PBDEs) | *Salmo salar* | Blood, liver, fillet, brain | CYP1A, VTG, ZR | Up-regulation of CYP1A, VTG and ZR gene | Boon et al. (2002) |
| 2,3,7,8-Tetrachlorodibenzo-*p*-dioxin (TCDD) | *Danio rerio* | Embryo | CYP1A | Hypoxia decreased TCDD induction of CYP1A mRNA, and decreased the potency of TCDD in causing edema | Prasch et al. (2004) |
| 4-Nonylphenol (4-NP) | *Sebastes schlegeli* | Liver | VTG | 4-NP disrupt the reproductive system of immature rockfish by suppressing the expression of vitellogenesis | Jung et al. (2006) |

**Table 2** Examples of studies using DDRT-PCR in fish toxicology

| Toxicants/ compounds | Species | Tissue | Target gene | Major findings | Reference |
|---|---|---|---|---|---|
| $E_2$, $EE_2$, diethyl-stilbestrol (DES) | *Cyprinodon variegatus* | Liver | 48 genes | $E_2$, $EE_2$ and DES act through the same pathway when administered through the water at low concentrations | Denslow et al. (2001a, b) |
| Cadmium (Cd) | *Chionodraco hamatus* | Liver | Seven genes | Up-regulation of gene coding for heat-shock protein HSP70 and down-regulation of transferrin gene | Carginale et al. (2002) |
| Effluents from pulp and paper mills | *Micropterus salmoides* | Liver | CYP1A | Decrease in reproductive success in females | Denslow et al. (2004) |
| Anthracene | *Fundulus heteroclitus* | Liver | 26 genes | CYP2N2 and two expressed sequence tags (15C1 and 18C2) act as good indicator of anthracene exposure | Peterson and Bain (2004) |
| Lindane | *Oreochromis niloticus* | Liver | 50 genes | Lindane triggers the expression of genes involved in immunological and stress responses | Colli-Dula et al. (2009) |
| Zearalenone (ZEA) | *Oncorhynchus mykiss* | Liver, ovary | 59 genes | ZEA interferes with blood coagulation and different cellular components | Woźny et al. (2012) |

offers opportunity of using high-quality genomic and transcriptomic data in non-model fish species.

It is evident that most of the researchers have used single compounds or simple mixtures to explore the relationship of gene expression and contaminant exposure in laboratory conditions. However, in natural environments, pollutants most often are present as complex mixtures rather than discrete chemicals. Therefore, future research should aim to understand the synergetic effects of multiple environmental stressors under field conditions. Furthermore, since chemical stressors invoke both toxic and adaptive processes, genetic variation of individual fish should also be taken into account for interpretation of alteration in gene expression profile

**Table 3** Examples of studies using suppression subtractive hybridization (SSH) in fish toxicology

| Toxicants/compounds | Species | Tissue | Target gene | Major findings | Reference |
|---|---|---|---|---|---|
| $EE_2$ | *Pleuronectes platessa* | Liver | 75 genes | Selective down-regulation of egg proteins, potentially diminishing the quality of eggs | Brown et al. (2004) |
| 2,3,7,8 Tetrachlorodibenzo-*p*-dioxin (TCDD) | *Oryzias latipes* | Brain, liver, and testis | 42 genes | Up-regulation of CYP1A mRNA in TCDD-exposed brain and liver | Volz et al. (2005) |
| Polychlorinated biphenyls (PCBs) | *Sebastiscus marmoratus* | Brain | 108 genes | PCBs cause differential expression of genes with function of neurotransmitter release, neuroendocrine, mitosis and cell proliferation, energy-related metabolism, general metabolism, signal protein, hemopoiesis system, immune system, and structure | Li et al. (2008) |
| Domestic effluents | *Dicentrarchus labrax* | Liver | CYP1A | Altered expression in genes involved in energy metabolism, immune system activity and antioxidant response | Nogueira et al. (2010) |
| Cadmium (Cd) | *Cyprinodon variegatus* | Larvae | 700 genes | Cd responsive genes are broadly involved in protective and repair functions as well as cellular and metabolic processes | Dangre et al. (2010) |
| Microcystin-LR (MCLR) | *Hypophthalmichthys molitrix* | Liver | 75 genes | Differential expression of genes with f functions of immune response, transport and cell metabolism | Qu et al. (2011) |
| Tetrodotoxin (TTX) | *Takifugu rubripes* | Liver | 1048 genes | High level of up-regulation in hepcidin precursor gene | Matsumoto et al. (2011) |

**Table 4** Examples of studies using reverse transcription quantitative polymerase chain reaction (RT-qPCR) in fish toxicology

| Toxicants/ compounds | Species | Tissue | Target gene | Major findings | Reference |
|---|---|---|---|---|---|
| $E_2$, α-zearalenol | *Oncorhynchus mykiss* | Liver | Zona Radiata (Zr), VTG | Induction of Zr and VTG | Celius et al. (2000) |
| $Cu^{2+}$, $Cd^{2+}$, $Hg^{2+}$, $Ni^{2+}$, $Pb^{2+}$, $Zn^{2+}$ | *Tilapia aurea_T. nilotica* | Liver, kidney, gill | MT | Metal contamination induce hepatic and renal MT mRNA levels | Cheung et al. (2004) |
| Bisphenol A, nonylphenol, octylphenol | *Kryptolebias marmoratus* | Brain, eye, gonad, intestine, liver, muscle, skin | MT | Down-regulation of metallothionein gene in both male and female individuals | Rhee et al. (2009) |
| Atrazine, chlorpyrifos | *Cyprinus carpio* | Brain, muscle | Acetylcholinesterase (AChE) | Down-regulation of AChE | Xing et al. (2010) |
| Estradiol, 2-methyl-testosterone | *Gobiocypris rarus* | Eye, brain, gill and gonads | *foxl2* | Up-regulation of *foxl2* with estradiol and down regulation with 2-methyl-testosterone | Jiang et al. (2011) |
| β-estradiol, 17-valerate (EV) | *Oryzias latipes* | Liver | Estrogen receptor α (ER-α) and vitellogenin-1 (VTG-1) | EV is a reproductive toxicant and cause reproductive failure in fish species | Lei et al. (2013) |
| Hg | *Danio rerio* | Larvae | P2X7R | Mercury cause toxicity by modulating P2X7R in zebrafish larvae | Cruz et al. (2013) |
| Cylindrospermopsin (CYN) | *Oreochromis niloticus* | Liver, kidney | GST | CYN cause hepatocyte and renal injury to fish | Gutiérrez-Praena et al. (2013) |
| Benzo[a]pyrene (BaP) | *Gobiocypris rarus* | Embryo | *cyp1a, 1b1, 1c1, gstm, ugt1a, abcb1, abcc1, abcc2, abcg2* | *cyp1a, 1b1, 1c1* and *abcc1, abcc2, abcg2* genes work in a coordinated manner to detoxify the effects of BaP in early developmental stage of fish | Yuan et al. (2014) |
| Nitrite | *Megalobrama amblycephala* | Liver | MaCAT, MaGPx1, MaCu/Zn-SOD | Imbalance between ROS and antioxidant defenses cause nitrite toxicity in *M. amblycephala* | Sun et al. (2014) |
| Tributyltin (TBT) | *Poecilia reticulata* | Brain | Aromatase | Suppression of aromatase cause the disturbance in reproductive behavior of fish by endocrine-disrupting action | Tian et al. (2015) |

**Table 5** Examples of studies using microarrays in fish toxicology

| Toxicants/compounds | Species | Tissue | Target gene | Major findings | Reference |
|---|---|---|---|---|---|
| Hypoxia | *Gillichthys mirabilis* | Liver, brain, skeletal, and cardiac muscle | 126 genes | Hypoxic environment first suppress the energy-requiring processes and then induce the genes needed for anaerobic ATP production and for gluconeogenesis | Gracey et al. (2001) |
| $E_2$, $EE_2$, DES, PNP, MXC, and ES | *Cyprinodon variegatus* | Liver | 30 genes | $E_2$, $EE_2$, DES, and MXC have similar genetic signatures | Larkin et al. (2003) |
| Cadmium | *Platichthys flesus* | Liver | CYP1A, Cu/Zn SOD, thioredoxin, a peroxiredoxin and a GST | Cadmium exposure cause oxidative stress in flounder | Sheader et al. (2006) |
| Methylmercury (MeHg) | *Pimephales promelas* | Liver | 76 genes in female and 42 in males | MeHg cause endocrine disruption in fish | Klaper et al. (2006) |
| Perfluorooctanoic acid (PFOA) | *Gobiocypris rarus* | Liver | 124 genes in male and 171 in female | Inhibition of thyroid hormone biosynthesis genes and induction of estrogen-responsive genes | Wei et al. (2008) |
| Perfluorooctane sulfonate (PFOS) | *Cyprinus carpio* | Liver | 1632 genes | PFOS affects the genes involved in energy metabolism, reproduction and stress response | Hagenaars et al. (2008) |
| PCBs, PAHs, Cu, and Cd | *Gadus morhua* | Liver, gill, and kidney | CYP1A, heme oxygenase, ferritin, and metallothionein | Microarrays can be used in diagnostic tool for environmental monitoring | Lie et al. (2009) |

| | | | | | |
|---|---|---|---|---|---|
| Dieldrin | *Largemouth bass* | Hypothalamus | 274 genes | Sub-chronic exposure to dieldrin alters the genes associated with cell metabolism, cell stability and integrity, stress, and DNA repair | Martyniuk et al. (2010) |
| Cu, Zn, and Cd | *Sparus aurata* | Fibroblast cell line (SAF1) | MT, GST, ATP7A | Increase in mRNA levels of MT and GST genes | Minghetti et al. (2011) |
| Titanium dioxide/$TiO_2$ and hydroxylated fullerenes/$C_{60}(OH)_{24}$ | *Danio rerio* | Embryo | 2693 genes | Nanoparticle exposure alters expression pattern in genes related to circadian rhythm, cell kinase activity, intracellular trafficking and immune response | Jovanovic et al. (2011) |
| $E_2$, nonylphenol (NP) and 2-chlorophenol (2CP) | *Oryzias latipes* | Liver | 1509 ESTs | Providing information of 1509 high-quality ESTs including 260 new EST sequences to O. latipes dbEST and GenBank of the NCBI | Pham et al. (2011) |
| Silver nanoparticles | *Oncorhynchus mykiss* | Liver | 207 genes | Toxicity of Ag differ depending on the presence of Ag nanoparticles and aggregates | Gagné et al. (2012) |
| Uranium | *Salmo salar* | Liver | 927 genes | Uranium exposure affects various toxicity pathways involved in mitochondrial functions, oxidative stress, nuclear receptor signaling, and organ damage | Song et al. (2014) |

**Table 6** Examples of studies using next-generation sequencing technology in fish toxicology

| Toxicants/compounds | Species | Tissue | Platform | Major findings | Reference |
|---|---|---|---|---|---|
| Cadmium (Cd) and Copper (Cu) | *Perca flavescens* | Liver | Roche 454 GS-FLX | Chronic exposure of Cd and Cu decrease in the transcription levels of genes involved in protein biosynthesis, immune system, and lipid and energy metabolism | Pierron et al. (2011) |
| Perfluorooctane sulfonate (PFOS) | *Oryzias melastigma* | Embryo | Illumina HiSeq 2000 | PFOS exposure cause neurobehavioral defects, mitochondrial dysfunction and the metabolism of proteins and fats | Huang et al. (2012) |
| 2,3,7,8-Tetrachlorodibenzo-*p*-dioxin (TCDD) | *Danio rerio* | Embryo | SOLiD | TCDD exposure causes modest changes in expression of miRNAs, including some (*miR-451, 23a, 23b, 24,* and *27e*) that are critical for hematopoiesis and cardiovascular development | Jenny et al. (2012) |
| Deepwater horizon oil | *Fundulus grandis* | Liver | Illumina GAIIx | DH oil exposure cause broad and complex genomic response including the expected AHR-mediated response, CYP genes, immune response, and choriogenin family genes | Garcia de la serrana et al. (2012) |
| Water-borne mixture of metals, including copper and zinc | *Salmo trutta* | Gills, liver and kidney | Illumina GAIIx | Metal- and ion-homeostasis pathways are likely to be the most important mechanisms contributing to the metal tolerance in brown trout | Webster et al. (2013) |
| Mercury (Hg) | *Brosme brosme* | Liver | Illumina HiSeq 2000 | Hg have toxic effects on lipid metabolism and oxidative stress | Olsvik et al. (2013) |
| Acidic aluminum-rich water | *Salmo salar* | Muscle | Illumina GAIIx | Identification of specific miRNAs as potential early markers for stress | Kure et al. (2013) |
| Oil sands process-affected water (OSPW) | *Pimephales promelas* | Liver | Illumina HiSeq 2000 | Exposure of OSPW alters gene expression of genes involved in oxidative metabolism, oxidative stress, apoptosis, and immune function | Wiseman et al. (2013) |
| Loading stress | *Coilia nasus* | Liver | Illumina HiSeq 2000 | Loading stress induce liver injury through the mitochondrial apoptosis pathway | Du et al. (2014) |
| Municipal wastewater | *Etheostoma caeruleum* | Ovary, testis | Roche 454 GS-FLX | Intermediate expression level of genes involved in sex differentiation (*sox9, foxl2* and *dmrt1*) and reproduction (*esr1, esrb, ar,vtg, cyp19a1,* and *cyp11a*) in intersex males compared to phenotypic males and females | Bahamonde et al. (2015) |

following chemical exposure. Although gene expression analyses provide the information about immediate toxic effects of pollutants and new hypotheses about their functions, it does not, as such, give functional information. Analysis of protein concentrations in a cell can give insights regarding gene function, interacting genes, and gene pathways. Therefore, multidisciplinary approaches should be implemented in monitoring the acute and chronic adverse effects of pollutant to understand the complexity of anthropogenic impacts on fish species.

**Conflict of Interest** The authors declare that they have no conflict of interest.

**Acknowledgment** The study was financially supported by the Ministry of Education, Youth and Sports of the Czech Republic—projects "CENAKVA" (No. CZ.1.05/2.1.00/01.0024) and "CENAKVA II" (No. LO1205 under the NPU I program).

## References

Adedeji OB, Durhan EJ, Garcia-Reyero N, Kahl MD, Jensen KM, Lalone CA, Makynen EA, Perkins EJ, Thomas L, Villeneuve DL, Ankley GT (2012) Short-term study investigating the estrogenic potency of diethylstilbesterol in the fathead minnow (*Pimephales promelas*). Environ Sci Technol 46:7826–7835

Alwine JC, Kemp DJ, Stark GR (1977) Method for detection of specific RNAs in agarose gels by transfer to diazobenzyloxymethyl-paper and hybridization with DNA probes. Proc Natl Acad Sci U S A 74:5350–5354

American Public Health Association, American Water Works Association, Water Pollution Control Federation (1960) Publication of "Bioassay methods for the evaluation of acute toxicity of industrial wastes and other substances to fish" in the 11th edition of Standard Methods. New York, p 626

Ankley GT, Villeneuve DL (2006) The fathead minnow in aquatic toxicology: past, present and future. Aquat Toxicol 78:91–102

Arini A, Gourves PY, Gonzalez P, Baudrimont M (2015) Metal detoxification and gene expression regulation after a Cd and Zn contamination: an experimental study on *Danio rerio*. Chemosphere 128:125–133

Arukwe A, Knudsen FR, Goksøyr A (1997) Fish zona radiata (eggshell) protein: a sensitive biomarker for environmental estrogens. Environ Health Perspect 105:418–422

Arukwe A, Kullman SW, Hinton DE (2001) Differential biomarker gene and protein expressions in nonylphenol and estradiol-17 beta treated juvenile rainbow trout (*Oncorhynchus mykiss*). Comp Biochem Physiol C Toxicol Pharmacol 129:1–10

Arukwe A, Kullman SW, Berg K, Goksoyr A, Hinton DE (2002) Molecular cloning of rainbow trout (*Oncorhynchus mykiss*) eggshell zona radiata protein complementary DNA: mRNA expression in 17beta-estradiol- and nonylphenol-treated fish. Comp Biochem Physiol B Biochem Mol Biol 132:315–326

Baer CE, Ippolito DL, Hussainzada N, Lewis JA, Jackson DA, Stallings JD (2014) Genome-wide gene expression profiling of acute metal exposures in male zebrafish. Genom Data 2:363–365

Bahamonde PA, McMaster ME, Servos MR, Martyniuk CJ, Munkittrick KR (2015) Molecular pathways associated with the intersex condition in rainbow darter (*Etheostoma caeruleum*) following exposures to municipal wastewater in the Grand River basin, ON, Canada. Part B. Aquat Toxicol 159:302–316

Boon JP, van Zanden JJ, Lewis WE, Zegers BN, Goksøyr A, Arukwe A (2002) The expression of CYP1A, vitellogenin and zona radiata proteins in Atlantic salmon (*Salmo salar*) after oral dosing with two commercial PBDE flame retardant mixtures: absence of short-term responses. Mar Environ Res 54:719–724

Bowman CJ, Kroll KJ, Hemmer MJ, Folmar LC, Denslow ND (2000) Estrogen-induced vitellogenin mRNA and protein in sheepshead minnow (*Cyprinodon variegatus*). Gen Comp Endocrinol 120:300–313

Bowman CJ, Kroll KJ, Gross TG, Denslow ND (2002) Estradiol-induced gene expression in largemouth bass (*Micropterus salmoides*). Mol Cell Endocrinol 196:67–77

Brazma A, Parkinson H, Sarkans U, Shojatalab M, Vilo J, Abeygunawardena N, Holloway E, Kapushesky M, Kemmeren P, Lara GG, Oezcimen A, Rocca-Serra P, Sansone S-A (2003) ArrayExpress-a public repository for microarray gene expression data at the EBI. Nucleic Acids Res 31:68–71

Brown M, Robinson C, Davies IM, Moffat CF, Redshaw J, Craft JA (2004) Temporal changes in gene expression in the liver of male plaice (*Pleuronectes platessa*) in response to exposure to ethynyl oestradiol analysed by macroarray and Real-Time PCR. Mutat Res 552:35–49

Buhler DR, Miranda CL, Henderson MC, Yang Y-H, Lee S-J, Wang-Buhler J-L (2000) Effects of 17β-estradiol and testosterone on hepatic mRNA/protein levels and catalytic activities of CYP2M1, CYP2K1, and CYP3A27 in rainbow trout (*Oncorhynchus mykiss*). Toxicol Appl Pharmacol 168:91–101

Cai Y, Li H, Hao L, Li G, Xie P, Chen J (2012) Identification of *cda* gene in bighead carp and its expression in response to microcystin-LR. Ecotoxicol Environ Saf 79:206–213

Camargo MM, Martinez CB (2007) Histopathology of gills, kidney and liver of a Neotropical fish caged in an urban stream. Neotrop Ichthyol 5:327–336

Carginale V, Capasso C, Scudiero R, Parisi E (2002) Identification of cadmium-sensitive genes in the Antarctic fish *Chionodraco hamatus* by messenger RNA differential display. Gene 299:117–124

Causton HC, Ren B, Koh SS, Harbison CT, Kanin E, Jennings EG, Lee TI, True HL, Lander ES, Young RA (2001) Remodeling of yeast genome expression in response to environmental changes. Mol Biol Cell 12:323–337

Celius T, Matthews JB, Giesy JP, Zacharewski TR (2000) Quantification of rainbow trout (*Oncorhynchus mykiss*) zona radiata and vitellogenin mRNA levels using real-time PCR after in vivo treatment with estradiol-17β or α-zearalenol. J Steroid Biochem Mol Biol 75:109–119

Chan KM, Ku LL, Chan PC-Y, Cheuk WK (2006) Metallothionein gene expression in zebrafish embryo-larvae and ZFL cell-line exposed to heavy metal ions. Mar Environ Res 62:S83–S87

Chen TT (1983) Identification and characterization of estrogen-responsive gene products in the liver of rainbow trout. Can J Biochem Cell Biol 61:802–810

Chen J (2015) Cardiac toxicity by sublethal 2,3,7,8-tetrachlorodibenzo-p-dioxin correlates with its anti-proliferation effect on cardiomyocytes in zebrafish embryos. Environ Toxicol Chem 34 (2):420–428

Cheung APL, Lam TH-J, Chan KM (2004) Regulation of Tilapia metallothionein gene expression by heavy metal ions. Mar Environ Res 58:389–394

Chow WS, Chan WK-L, Chan KM (2011) Toxicity assessment and vitellogenin expression in zebrafish (*Danio rerio*) embryos and larvae acutely exposed to bisphenol A, endosulfan, heptachlor, methoxychlor and tetrabromobisphenol A. J Appl Toxicol 33:670–678

Colli-Dula R, Zuniga-Aguilar J-J, Albores-Medina A, Zapata-Perez O (2009) Identification of genes expressed as a result of lindane exposure in *Oreochromis niloticus* using differential display. Ecotoxicol Environ Saf 72:1406–1412

Conti GO, Copat C, Ledda C, Fiore M, Fallico R, Sciacca S, Ferrante M (2012) Evaluation of heavy metals and polycyclicaromatichydrocarbons (PAHs) in *Mullus barbatus* from Sicily Channel and risk-based consumption limits. Bull Environ Contam Toxicol 88:946–950

Copat C, Bella F, Castaing M, Fallico R, Sciacca S, Ferrante M (2012a) Heavy metals concentrations in fish from Sicily (Mediterranean Sea) and evaluation of possible health risks to consumers. Bull Environ Contam Toxicol 88:78–83

Copat C, Maggiore R, Arena G, Lanzafame S, Fallico R, Sciacca S, Ferrante M (2012b) Evaluation of a temporal trend heavy metals contamination in *Posidonia oceanica* (L.) Delile, (1813) along the western coastline of Sicily (Italy). J Environ Monit 14:187–192

Cruz FF, Leite CE, Pereira TCB et al (2013) Assessment of mercury chloride-induced toxicity and the relevance of P2X7 receptor activation in zebrafish larvae. Comp Biochem Physiol C Toxicol Pharmacol 158:159–164

Dangre AJ, Manning S, Brouwer M (2010) Effects of cadmium on hypoxia-induced expression of hemoglobin and erythropoietin in larval sheepshead minnow, *Cyprinodon variegatus*. Aquat Toxicol 99:168–175

Davis AP, Grondin CJ, Lennon-Hopkins K, Saraceni-Richards C, Sciaky D, King BL, Wiegers TC, Mattingly CJ (2015) The Comparative Toxicogenomics Database's 10th year anniversary: update 2015. Nucleic Acids Res 43:D914–D920

Denslow ND, Chow M, Chow MM, Bonomelli S, Folmar LC, Heppell SA, Sullivan SV (1997a) Development of biomarkers for environmental contaminants affecting fish. In: Rolland RM, Gilbertson M, Peterson RE (eds) Chemically induced alterations in functional development and reproduction of fishes , Proceeding from a session at the 1995 Wingspread Conference, Society of Environmental Toxicology and Chemistry (SETAQ). Publication Series. SETAC Press, Pensacola, pp 73–86

Denslow ND, Chow MM, Folmar LC, Bonomelli SL, Heppell SA, Sullivan CV (1997b) Development of antibodies to teleost vitellogenins: potential biomarkers for environmental estrogens. In: Bengston DA, Hen-schel DS (eds) Environmental toxicology and risk assessment: biomarkers and risk assessment, vol 5. American Society for Testing and Materials, Philadelphia, pp 23–36, ASTM STP 1306

Denslow ND, Bowman CJ, Ferguson RJ, Lee HS, Hemmer MJ, Folmar LC (2001a) Induction of gene expression in sheepshead minnows (*Cyprinodon variegatus*) treated with 17 beta-estradiol, diethylstilbesterol, or ethinylestradiol: the use of mRNA fingerprints as an indicator of gene regulation. Gen Comp Endocrinol 121:250–260

Denslow ND, Lee HS, Bowman CJ, Hemmer MJ, Folmar LC (2001b) Multiple responses in gene expression in fish treated with estrogen. Comp Biochem Physiol B Biochem Mol Biol 129:277–282

Denslow ND, Kocerha J, Sepulveda MS, Gross T, Holm SE (2004) Gene expression fingerprints of largemouth bass (*Micropterus salmoides*) exposed to pulp and paper mill effluents. Mutat Res 552:19–34

Diatchenko L, Lau YF, Campbell AP et al (1996) Suppression subtractive hybridization: a method for generating differentially regulated or tissue-specific cDNA probes and libraries. Proc Natl Acad Sci U S A 93:6025–6030

Dixon TJ, Taggart JB, George SG (2002) Application of real time PCR determination to assess inter-animal variabilities in CYP1A induction in the European flounder (*Platichthys flesus*). Mar Environ Res 54:267–270

Du F, Xu G, Nie Z, Xu P, Gu R (2014) Transcriptome analysis gene expression in the liver of *Coilia nasus* during the stress response. BMC Genomics 15:558

Du Z, Wang G, Gao S, Wang Z (2015) Aryl organophosphate flame retardants induced cardiotoxicity during zebrafish embryogenesis: by disturbing expression of the transcriptional regulators. Aquat Toxicol 161:25–32

Edgar R, Domrachev M, Lash AE (2002) Gene Expression Omnibus: NCBI gene expression and hybridization array data repository. Nucleic Acids Res 30:207–210

Fan C-Y, Cowden J, Simmons SO, Padilla S, Ramabhadran R (2010) Gene expression changes in developing zebrafish as potential markers for rapid developmental neurotoxicity screening. Neurotoxicol Teratol 32:91–98

Ferraris M, Flora A, Chiesara E, Fornasari D, Lucchetti H, Marabini L, Frigerio S, Radice S (2005) Molecular mechanism of the aryl hydrocarbon receptor activation by the fungicide iprodione in rainbow trout (*Oncorhynchus mykiss*) hepatocytes. Aquat Toxicol 72:209–220

Flouriot G, Pakdel F, Ducouret B, Ledrean Y, Valotaire Y (1997) Differential regulation of two genes implicated in fish reproduction: vitellogenin and estrogen receptor genes. Mol Reprod Dev 48:317–323

Folmar LC, Denslow ND, Rao V, Chow M, Crain DA, Enblom J, Marcino J, Guillette LJ Jr (1996) Vitellogenin induction and reduced serum testosterone concentrations in feral male carp, *Cyprinus carpio*. Captured near a major metropolitan sewage treatment plant. Environ Health Perspect 104:1096–1101

Funkenstein B, Bowman C, Denslow ND, Cardinali M, Carnevali O (2000) Contrasting effects of estrogen on transthyretin and vitellogenin expression in males of the marine fish, *Sparus aurata*. Mol Cell Endocrinol 167:33–41

Funkenstein B (2001) Developmental expression, tissue distribution and hormonal regulation of fish (*Sparus aurata*) serum retinol-binding protein. Comp Biochem Physiol B Biochem Mol Biol 129:613–622

Gagné F, André C, Skirrow R, Gélinas M, Auclair J, van Aggelen G, Turcotte P, Gagnon C (2012) Toxicity of silver nanoparticles to rainbow trout: a toxicogenomic approach. Chemosphere 89:615–622

Garcia de la serrana D, Estévez A, Andree K, Johnston IA (2012) Fast skeletal muscle transcriptome of the Gilthead sea bream (*Sparus aurata*) determined by next generation sequencing. BMC Genomics 13:181

Garcia-Reyero N, Denslow ND (2006) Applications of genomic technologies to the study of organochlorine pesticide-induced reproductive toxicity in fish. J Pestic Sci 31(3):252–262

Garcia-Reyero N, Griffitt RJ, Liu L, Kroll KJ, Farmerie WG, Barber DS, Denslow ND (2008) Construction of a robust microarray from a non-model species (largemouth bass) using pyrosequencing technology. J Fish Biol 72:2354–2376

Geoghegan F, Katsiadaki I, Williams TD, Chipman JK (2008) A cDNA microarray for the three-spined stickleback, *Gasterosteus aculeatus* L, and analysis of the interactive effects of oestradiol and dibenzanthracene exposures. J Fish Biol 72:2133–2153

Glisic B, Hrubik J, Fa S, Dopudj N, Kovacevic R, Andric N (2016) Transcriptional profiles of glutathione-S transferase isoforms, Cyp, and AOE genes in atrazine-exposed zebrafish embryos. Environ Toxicol 31:233–244

Goodale BC, Tilton SC, Corvi MM, Wilson GR, Janszen DB, Anderson KA, Waters KM, Tanguay RL (2013) Structurally distinct polycyclic aromatic hydrocarbons induce differential transcriptional responses in developing zebrafish. Toxicol Appl Pharmacol 272:656–670

Gracey AY, Troll JV, Somero GN (2001) Hypoxia-induced gene expression profiling in the euryoxic fish *Gillichthys mirabilis*. Proc Natl Acad Sci U S A 98:1993–1998

Griffitt RJ, Hyndman K, Denslow ND, Barber DS (2009) Comparison of molecular and histological changes in zebrafish gills exposed to metallic nanoparticles. Toxicol Sci 107:404–415

Gust KA, Brasfield SM, Stanley JK, Wilbanks MS, Chappell P, Perkins EJ, Lotufo GR, Lance RF (2011) Genomic investigation of year-long and multigenerational exposures of fathead minnow to the munitions compound RDX. Environ Toxicol Chem 30:1852–1864

Gutiérrez-Praena D, Jos A, Pichardo S, Puerto M, Cameán AM (2013) Influence of the exposure way and the time of sacrifice on the effects induced by a single dose of pure Cylindrospermopsin on the activity and transcription of glutathione peroxidase and glutathione-S-transferase enzymes in Tilapia (*Oreochromis niloticus*). Chemosphere 90:986–992

Hagenaars A, Knapen D, Meyer IJ, van der Ven K, Hoff P, De Coen W (2008) Toxicity evaluation of perfluorooctane sulfonate (PFOS) in the liver of common carp (*Cyprinus carpio*). Aquat Toxicol 88:155–163

Hermsen SAB, Pronk TE, van den Brandhof E-J, van der Ven LTM, Piersma AH (2012) Concentration-response analysis of differential gene expression in the zebrafish embryo toxicity test following flusilazole exposure. Toxicol Sci 127(1):303–312
Higuchi R, Fockler C, Dollinger G, Watson R (1993) Kinetic PCR analysis: real-time monitoring of DNA amplification reactions. Biotechnology 11:1026–1030
Hofsteen P, Mehta V, Kim M-S, Peterson RE, Heideman W (2013) TCDD inhibits heart regeneration in adult zebrafish. Toxicol Sci 132(1):211–221
Hogan NS, Wartman CA, Finley MA, van der Lee JG, van den Heuvel MR (2008) Simultaneous determination of androgenic and estrogenic endpoints in the threespine stickleback (*Gasterosteus aculeatus*) using quantitative RT-PCR. Aquat Toxicol 90:269–276
Hong HN, Kim HN, Park KS, Lee S-K, Gu MB (2007) Analysis of the effects diclofenac has on Japanese medaka (*Oryzias latipes*) using real-time PCR. Chemosphere 67:2115–2121
Hook SE, Skillman AD, Small JA, Schultz IR (2006) Gene expression patterns in rainbow trout, *Oncorhynchus mykiss*, exposed to a suite of model toxicants. Aquat Toxicol 77:372–385
Huang Q, Dong S, Fang C, Wu X, Ye T, Lin Y (2012) Deep sequencing-based transcriptome profiling analysis of *Oryzias melastigma* exposed to PFOS. Aquat Toxicol 120–121:54–58
Hunn JB (1989) History of acute toxicity tests with fish, 1863–1987. In: Investigations in fish control 98. US Fish and Wildlife Service, LaCrosse, WI, pp 1–10
Ikeo K, Ishi-i J, Tamura T, Gojobori T, Tateno Y (2003) CIBEX: Center for Information Biology gene EXpression database. C R Biol 326:1079–1082
Jayasundara N, Garner LV, Meyer JN, Erwin KN, Di Giulio RT (2015) AHR2-mediated transcriptomic responses underlying the synergistic cardiac developmental toxicity of PAHs. Toxicol Sci 143(2):469–481
Jenny MJ, Aluru N, Hahn ME (2012) Effects of short-term exposure to 2,3,7,8-tetrachlorodibenzo-p-dioxin on microRNA expression in zebrafish embryos. Toxicol Appl Pharmacol 264:262–273
Jiang W, Yang Y, Zhao D, Liu X, Duan J, Xie S, Zhao H (2011) Effects of sexual steroids on the expression of *foxl2* in *Gobiocypris rarus*. Comp Biochem Physiol B Biochem Mol Biol 160:187–193
Jing J, Liu H, Chen H, Hu S, Xiao K, Ma X (2013) Acute effect of copper and cadmium exposure on the expression of heat shock protein 70 in the Cyprinidae fish *Tanichthys albonubes*. Chemosphere 91:1113–1122
Jobling S, Sheahan D, Osborne JA, Matthiessen P, Sumpter JP (1996) Inhibition of testicular growth in rainbow trout (*Oncorhynchus mykiss*) exposed to estrogenic alkylphenolic chemicals. Environ Toxicol Chem 15:194–202
Jovanovic B, Ji T, Palic D (2011) Gene expression of zebrafish embryos exposed to titanium dioxide nanoparticles and hydroxylated fullerenes. Ecotoxicol Environ Saf 74:1518–1525
Ju Z, Wells MC, Walter RB (2007) DNA microarray technology in toxicogenomics of aquatic models: methods and applications. Comp Biochem Physiol C Toxicol Pharmacol 145:5–14
Jung J-H, Shim W, Addison RF, Baek J, Han C-H (2006) Protein and gene expression of VTG in response to 4-nonylphenol in rockfish (*Sebastes schlegeli*). Comp Biochem Physiol C Toxicol Pharmacol 143:162–170
Klaper R, Rees CB, Drevnick P, Weber D, Sandheinrich M, Carvan MJ (2006) Gene expression changes related to endocrine function and decline in reproduction in fathead minnow (*Pimephales promelas*) after dietary methylmercury exposure. Environ Health Perspect 114 (9):1337–1343
Klaper R, Carter BJ, Richter CA, Drevnick PE, Sandheinrich MB, Tillitt DE (2008) Use of a 15 k gene microarray to determine gene expression changes in response to acute and chronic methylmercury exposure in the fathead minnow *Pimephales promelas* Rafinesque. J Fish Biol 72:2207–2280
Knoebl I, Blum JL, Hemmer MJ, Denslow ND (2006) Temporal gene induction patterns in sheepshead minnows exposed to 17β-estradiol. J Exp Zool 305A:707–719

Kriegsfeld LJ, Mei DF, Bentley GE, Ubuka T, Mason AO, Inoue K, Ukena K, Tsutsui K, Silver R (2006) Identification and characterization of a gonadotropin-inhibitory system in the brains of mammals. Proc Natl Acad Sci U S A 103:2410–2415

Kure EH, Sæbø M, Stangeland AM, Hamfjord J, Hytterød S, Heggenes J, Lydersen E (2013) Molecular responses to toxicological stressors: profiling microRNAs in wild Atlantic salmon (*Salmo salar*) exposed to acidic aluminum-rich water. Aquat Toxicol 138–139:98–104

Lam SH, Hlaing MM, Zhang X, Yan C, Duan Z, Zhu L, Ung CY, Mathavan S, Ong CN, Gong Z (2011) Toxicogenomic and phenotypic analyses of bisphenol-A early-life exposure toxicity in zebrafish. PLoS One 6(12):e28273

Larkin P, Folmar LC, Hemmer MJ, Poston AJ, Denslow ND (2003) Expression profiling of estrogenic compounds using a sheepshead minnow cDNA macroarray. Environ Health Perspect 111:839–846

Lazier CB, Langley S, Ramsey NB, Wright JM (1996) Androgen inhibition of vitellogenin gene expression in tilapia (*Oreochromis niloticus*). Gen Comp Endocrinol 104:321–329

Leaver MJ, George SG (2000) A cytochrome P4501B gene from a fish, *Pleuronectes platessa*. Gene 256:83–91

Lefebvre KA, Tilton SC, Bammler TK, Beyer RP, Srinouanprachan S, Stapleton PL, Farin FM, Gallagher EP (2009) Gene expression profiles in zebrafish brain after acute exposure to domoic acid at symptomatic and asymptomatic doses. Toxicol Sci 107(1):65–77

Lei B, Kang J, Yu Y, Zha J, Li W, Wang Z (2013) β-estradiol 17-valerate affects embryonic development and sexual differentiation in Japanese medaka (*Oryzias latipes*). Aquat Toxicol 134–135:128–134

Lerebours A, Bourdineaud JP, van der Ven K, Vandenbrouck T, Gonzalez P, Camilleri V, Floriani M, Garnier-Laplace J, Adam-Guillermin C (2010) Sublethal effects of waterborne uranium exposures on the zebrafish brain: transcriptional responses and alterations of the olfactory bulb ultrastructure. Environ Sci Technol 44:1438–1443

Li B, Wang C, Ye K, Yu A, Chen Y, Zuo Z (2008) Differential gene expression in the brain of *Sebastiscus marmoratus* in response to exposure to polychlorinated biphenyls (PCBs). Mar Environ Res 66:548–552

Li X, Liu L, Zhang Y, Fang Q, Li Y, Li Y (2013a) Toxic effects of chlorpyrifos on lysozyme activities, the contents of complement C3 and IgM, and IgM and complement C3 expressions in common carp (*Cyprinus carpio* L.). Chemosphere 93:428–433

Li X, Ma J, Fang Q, Li Y (2013b) Transcription alterations of microRNAs, cytochrome P4501A1 and 3A65, and AhR and PXR in the liver of zebrafish exposed to crude microcystins. Toxicon 73:17–22

Liang P, Pardee A (1992) Differential display using one base anchor oligo-dT primers. Nucleic Acids Res 22:5763–5764

Lie KK, Lanzen A, Breilid H, Olsvik PA (2009) Gene expression profiling in Atlantic cod (*Gadus morhua* L.) from two contaminated sites using a custom-made cDNA microarray. Environ Toxicol Chem 28:1711–1721

Liu Q, Spitsbergen JM, Cariou R, Huang CY, Jiang N, Goetz G, Hutz RJ, Tonellato PJ, Carvan MJ 3rd (2014) Histopathologic alterations associated with global gene expression due to chronic dietary TCDD exposure in juvenile zebrafish. PLoS One 9(7):e100910

Luzio A, Monteiro SM, Fontaínhas-Fernandes AA, Pinto-Carnide O, Matos M, Coimbra AM (2013) Copper induced upregulation of apoptosis related genes in zebrafish (*Danio rerio*) gill. Aquat Toxicol 128–129:183–189

Martinovic D, Villeneuve DL, Kahl MD, Blake LS, Brodin JD, Ankley GT (2009) Hypoxia alters gene expression in the gonads of zebrafish (*Danio rerio*). Aquat Toxicol 95:258–272

Martyniuk CJ, Gerrie ER, Popesku JT, Ekker M, Trudeau VL (2007) Microarray analysis in the zebrafish (*Danio rerio*) liver and telencephalon after exposure to low concentration of 17alpha-ethinylestradiol. Aquat Toxicol 84:38–49

Martyniuk CJ, Kroll KJ, Doperalski NJ, Barber DS, Denslow ND (2010) Genomic and proteomic responses to environmentally relevant exposures to dieldrin: indicators of neurodegeneration? Toxicol Sci 117(1):190–199

Matsumoto T, Ishizaki S, Nagashima Y (2011) Differential gene expression profile in the liver of the marine puffer fish *Takifugu rubripes* induced by intramuscular administration of tetrodotoxin. Toxicon 57:304–310

Mattingly CJ, Colby GT, Forrest JN, Boyer JL (2003) The Comparative Toxicogenomics Database (CTD). Environ Health Perspect 111:793–795

Mellanen P, Soimasuo M, Holmbom B, Oikari A, Santti R (1999) Expression of the vitellogenin gene in the liver of juvenile whitefish (*Coregonus lavaretus* L. *s.l.*) exposed to effluents from pulp and paper mills. Ecotoxcol Environ Saf 43:133–137

Meyer JN, Volz DC, Freedman JH, Di Giulio RT (2005) Differential display of hepatic mRNA from killifish (*Fundulus heteroclitus*) inhabiting a superfund estuary. Aquat Toxicol 73:327–341

Minghetti M, Leaver MJ, Taggart JB, Casadei E, Auslander M, Tom M, George SG (2011) Copper induces Cu-ATPase ATP7A mRNA in a fish cell line, SAF1. Comp Biochem Physiol C Toxicol Pharmacol 154:93–99

Moens LN, Soetaert A, van der Ven K, Del-Favero J, De Coen WM (2007a) Use of suppression subtractive hybridization PCR for the development of cDNA arrays for the detection of endocrine disruption in carp (*Cyprinus carpio*). Comp Biochem Physiol D Genomics Proteomics 2:18–33

Moens LN, Smolders R, van der Ven K, van Remortel P, Del-Favero J, DeCoen WM (2007b) Effluent impact assessment using microarray-based analysis in common carp: a systems toxicology approach. Chemosphere 67:2293–2304

Murata K, Sugiyama H, Yasumasu S, Iuchi I, Yasumasu I, Yamagami K (1997) Cloning of cDNA and estrogen induced hepatic gene expression for choriogenin H, a precursor protein of the fish egg envelope (chorion). Proc Natl Acad Sci U S A 94:2050–2055

Nogueira P, Pacheco M, Pereira ML, Mendo S, Rotchell JM (2010) Anchoring novel molecular biomarker responses to traditional responses in fish exposed to environmental contamination. Environ Pollut 158:1783–1790

Olsvik PA, Lindgren M, Maage A (2013) Mercury contamination in deep-water fish: transcriptional responses in tusk (*Brosme brosme*) from a fjord gradient. Aquat Toxicol 144–145:172–185

Oppen-Berntsen DO, Arukwe A, Yadetie F, Lorens JB, Male R (1999) Salmon eggshell protein expression: a marker for environmental estrogens. Mar Biotechnol 1:252–260

Penny C, Adams C (1863) Fourth report, Royal Commission on Pollution of Rivers in Scotland. Vol. 2. Evidence. London (*in* Jones 1964), pp 377–391

Peterson JSK, Bain LJ (2004) Differential gene expression in anthracene-exposed mummichogs (*Fundulus heteroclitus*). Aquat Toxicol 66:345–355

Peterson RE, Theobald HM, Kimmel GL (1993) Developmental and reproductive toxicity of dioxins and related compounds: cross-species comparisons. Crit Rev Toxicol 23:283–335

Pham CH, Park KS, Kim BC, Kim HN, Gu MB (2011) Construction and characterization of Japanese medaka (*Oryzias latipes*) hepatic cDNA library and its implementation to biomarker screening in aquatic toxicology. Aquat Toxicol 105:569–575

Pham CH, Yi J, Gu MB (2012) Biomarker gene response in male Medaka (*Oryzias latipes*) chronically exposed to silver nanoparticle. Ecotoxicol Environ Saf 78:239–245

Pierron F, Normandeau E, Defo MA, Campbell PGC, Bernatchez L, Couture P (2011) Effects of chronic metal exposure on wild fish populations revealed by high-throughput cDNA sequencing. Ecotoxicology 20:1388–1399

Porcher C, Malinge MC, Picat C, Grandchamp B (1992) A simplified method for determination of specific DNA or RNA copy number using quantitative PCR and an automatic DNA sequencer. Biotechniques 13:106–114

Prasch AL, Andreasen EA, Peterson RE, Heideman W (2004) Interactions between 2,3,7,8-tetrachlorodibenzo-*p*-dioxin (TCDD) and hypoxia signaling pathways in zebrafish: hypoxia decreases responses to TCDD in zebrafish embryos. Toxicol Sci 78:68–77

Qu X, Zhang K, Cui Z, Zhang Y, Jiang J, Feng L, Liu Q (2011) Construction and analysis of liver suppression subtractive hybridization library of silver carp (*Hypophthalmichthys molitrix*) intraperitoneally injected with microcystin-LR. Aquat Toxicol 105:151–156

Ramsay G (1998) DNA chips: state-of-the art. Nat Biotechnol 16:40–44

Rhee J-S, Raisuddin S, Hwang D-S, Lee K-W, Kim IC, Lee J-S (2009) Differential expression of metallothionein (MT) gene by trace metals and endocrine-disrupting chemicals in the hermaphroditic mangrove killifish, *Kryptolebias marmoratus*. Ecotoxicol Environ Saf 72:206–212

Roling JA, Bain LJ, Baldwin WS (2004) Differential gene expression in mummichogs (*Fundulus heteroclitus*) following treatment with pyrene: comparison to a creosote contaminated site. Mar Environ Res 57:377–395

Sammar M, Babin PJ, Durliat M, Meiri I, Zchori I, Elizur A, Lubzens E (2001) Retinol binding protein in rainbow trout: molecular properties and mRNA expression in tissues. Gen Comp Endocrinol 123:51–61

Santos D, Matos M, Coimbra AM (2014) Developmental toxicity of endocrine disruptors in early life stages of zebrafish, a genetic and embryogenesis study. Neurotoxicol Teratol 46:18–25

Sasaki YF, Izumiyama F, Nishidate E, Ishibashi S, Tsuda S, Matsusaka N, Asano N, Saotome K, Sofuni T, Hayashi M (1997) Detection of genotoxicity of polluted sea water using shellfish and the alkaline single-cell gel electrophoresis (SCE) assay: a preliminary study. Mutat Res 393:133–139

Sawle AD, Wit E, Whale G, Cossins AR (2010) An information-rich alternative, chemicals testing strategy using a high definition toxicogenomics and zebrafish (*Danio rerio*) embryos. Toxicol Sci 118(1):128–139

Schiller V, Zhang X, Hecker M, Schäfers C, Fischer R, Fenske M (2014) Species-specific considerations in using the fish embryo test as an alternative to identify endocrine disruption. Aquat Toxicol 155:62–72

Sheader DL, Williams TD, Lyons BP, Chipman JK (2006) Oxidative stress response of European flounder (*Platichthys flesus*) to cadmium determined by a custom cDNA microarray. Mar Environ Res 62:33–44

Shumway M, Cochrane G, Sugawara H (2010) Archiving next generation sequencing data. Nucleic Acids Res 38:D870–D871

Song Y, Salbu B, Heier LS, Teien HC, Lind OC, Oughton D, Petersen K, Rosseland BO, Skipperud L, Tollefsen KE (2012) Early stress responses in Atlantic salmon (*Salmo salar*) exposed to environmentally relevant concentrations of uranium. Aquat Toxicol 112:62–71

Song Y, Salbu B, Teien H-C, Heier LS, Rosseland BO, Høgåsen T, Tollefsen KE (2014) Hepatic transcriptomic profiling reveals early toxicological mechanisms of uranium in Atlantic salmon (*Salmo salar*). BMC Genomics 15:694

Sumpter JP, Jobling S (1995) Vitellogenin as a biomarker for estrogenic contamination of the aquatic environment. Environ Health Perspect 103:173–178

Sun S, Ge X, Zhu J, Xuan F, Jiang X (2014) Identification and mRNA expression of antioxidant enzyme genes associated with the oxidative stress response in the Wuchang bream (*Megalobrama amblycephala* Yih) in response to acute nitrite exposure. Comp Biochem Physiol C Toxicol Pharmacol 159:69–77

Tian H, Wu P, Wang W, Ru S (2015) Disruptions in aromatase expression in the brain, reproductive behavior, and secondary sexual characteristics in male guppies (*Poecilia reticulata*) induced by tributyltin. Aquat Toxicol 162:117–125

Tilton SC, Gerwick LG, Hendricks JD et al (2005) Use of a rainbow trout oligonucleotide microarray to determine transcriptional patterns in aflatoxin B1-induced hepatocellular carcinoma compared to adjacent liver. Toxicol Sci 88(2):319–330

Turner C, Sawle A, Fenske M, Cossins A (2012) Implications of the solvent vehicles dimethylformamide and dimethylsulfoxide for establishing transcriptomic endpoints in the zebrafish embryo toxicity test. Environ Toxicol Chem 31:593–604
van Boxtel AL, Pieterse B, Cenijn P, Kamstra JH, Brouwer A, van Wieringen W, de Boer J, Legler J (2010) Dithiocarbamates induce craniofacial abnormalities and downregulate *sox9a* during zebrafish development. Toxicol Sci 117(1):209–217
Varela-Valencia R, Gomez-Ortiz N, Oskam G, de Coss R, Rubio-Pina J, Rio-Garcıa M, Albores-Medina A, Zapata-Perez O (2014) The effect of titanium dioxide nanoparticles on antioxidant gene expression in tilapia (*Oreochromis niloticus*). J Nanopart Res 16:2369
Volz DC, Bencic DC, Hinton DE, Law JM, Kullman SW (2005) 2,3,7,8-Tetrachlorodibenzo-p-dioxin (TCDD) induces organ-specific differential gene expression in male Japanese medaka (*Oryzias latipes*). Toxicol Sci 85:572–584
Volz DC, Hinton DE, Law JM, Kullman SW (2006) Dynamic gene expression changes precede dioxin-induced liver pathogenesis in medaka fish. Toxicol Sci 89(2):524–534
Wang Z, Gerstein M, Snyder M (2009) RNA-Seq: a revolutionary tool for transcriptomics. Nat Rev Genet 10(1):57–63
Wang R-L, Bencic D, Lazorchak J, Villeneuve D, Ankley GT (2011) Transcriptional regulatory dynamics of the hypothalamic–pituitary–gonadal axis and its peripheral pathways as impacted by the 3-beta HSD inhibitor trilostane in zebrafish (*Danio rerio*). Ecotoxicol Environ Saf 74:1461–1470
Wang Q, Lam JCW, Han J, Wang X, Guo Y, Lam PKS, Zhou B (2015a) Developmental exposure to the organophosphorus flame retardant tris (1,3-dichloro-2-propyl) phosphate: estrogenic activity, endocrine disruption and reproductive effects on zebrafish. Aquat Toxicol 160:163–171
Wang Z, Zhang X, Tian H, Wang W, Ru S (2015b) Effects of monocrotophos pesticide on steroidogenesis and transcription of steroidogenic enzymes in rainbow trout RTG-2 cells involving the protein kinase A signal pathway. Toxicol In Vitro 29:155–161
Waring JF, Ciurlionis R, Jolly RA, Heindel M, Ulrich RG (2001) Microarray analysis of hepatotoxins in vitro reveals a correlation between gene expression profiles and mechanisms of toxicity. Toxicol Lett 120:359–368
Webster TMU, Bury N, van Aerle R, Santos EM (2013) Global transcriptome profiling reveals molecular mechanisms of metal tolerance in a chronically exposed wild population of brown trout. Environ Sci Technol 47:8869–8877
Wei Y, Liu Y, Wang J, Tao Y, Dai J (2008) Toxicogenomic analysis of the hepatic effects of perfluorooctanoic acid on rare minnows (*Gobiocypris rarus*). Toxicol Appl Pharmacol 226:285–297
Wei Y, Shi X, Zhang H, Wang J, Zhou B, Dai J (2009) Combined effects of polyfluorinated and perfluorinated compounds on primary cultured hepatocytes from rare minnow (*Gobiocypris rarus*) using toxicogenomic analysis. Aquat Toxicol 95:27–36
Williams DE, Lech JJ, Buhler DR (1998) Xenobiotics and xenoestrogens in fish: modulation of cytochrome P450 and carcinogenesis. Mutat Res 399:179–192
Williams TD, Gensberg K, Minchin SD, Chipman JK (2003) A DNA expression array to detect toxic stress response in European flounder (*Platichthys flesus*). Aquat Toxicol 65:141–157
Williams TD, Diab AM, Gubbins M, Collins C, Matejusova I, Kerr R, Chipman JK, Kuiper R, Vethaak AD, George SG (2013) Transcriptomic responses of European flounder (*Platichthys flesus*) liver to a brominated flame retardant mixture. Aquat Toxicol 142–143:45–52
Williams TD, Mirbahai L, Chipman JK (2014) The toxicological application of transcriptomics and epigenomics in zebrafish and other teleosts. Brief Funct Genomics 13(2):157–171
Wintz H, Yoo LJ, Loguinov A, Wu YY, Steevens JA, Holland RD, Beger RD, Perkins EJ, Hughes O, Vulpe CD (2006) Gene expression profiles in fathead minnow exposed to 2,4-DNT: correlation with toxicity in mammals. Toxicol Sci 94(1):71–82

Wiseman SB, He Y, Din MG, Martin JW, Jones PD, Hecker M, John P, Giesy JP (2013) Transcriptional responses of male fathead minnows exposed to oil sands process-affected water. Comp Biochem Physiol C Toxicol Pharmacol 157:227–235

Woo S, Won H, Lee A, Yum S (2012) Oxidative stress and gene expression in diverse tissues of *Oryzias javanicus* exposed to 17β-estradiol. Mol Cell Toxicol 8:263–269

Woźny M, Brzuzan P, Wolińska L, Góra M, Łuczyński MK (2012) Differential gene expression in rainbow trout (*Oncorhynchus mykiss*) liver and ovary after exposure to zearalenone. Comp Biochem Physiol C Toxicol Pharmacol 156:221–228

Wu M, Zuo Z, Li B, Huang L, Chen M, Wang C (2013) Effects of low-level hexabromocyclododecane (HBCD) exposure on cardiac development in zebrafish embryos. Ecotoxicology 22:1200–1207

Xing H, Han Y, Li S, Wang J, Wang X, Xu S (2010) Alterations in mRNA expression of acetylcholinesterase in brain and muscle of common carp exposed to atrazine and chlorpyrifos. Ecotoxicol Environ Saf 73:1666–1670

Yadetie F, Male R (2002) Effects of 4-nonylphenol on gene expression of pituitary hormones in juvenile Atlantic salmon (*Salmo salar*). Aquat Toxicol 58:113–129

Yadetie F, Karlsen OA, Lanzén A, Berg K, Olsvik P, Hogstrand C, Goksøyr A (2013) Global transcriptome analysis of Atlantic cod (*Gadus morhua*) liver after in vivo methylmercury exposure suggests effects on energy metabolism pathways. Aquat Toxicol 126:314–325

Yamauchi M, Kim E-Y, Iwata H, Shima Y, Tanabe S (2006) Toxic effects of 2,3,7,8-tetrachlorodibenzo-*p*-dioxin (TCDD) in developing red seabream (*Pagrus major*) embryo: an association of morphological deformities with AHR1, AHR2 and CYP1A expressions. Aquat Toxicol 80:166–179

Yamauchi R, Ishibashi H, Hirano M, Mori T, Kim J-W, Arizono K (2008) Effects of synthetic polycyclic musks on estrogen receptor, vitellogenin, pregnane X receptor, and cytochrome P450 3A gene expression in the livers of male medaka (*Oryzias latipes*). Aquat Toxicol 90:261–268

Yan W, Zhou Y, Yang J, Li S, Hu D, Wang J, Chen J, Li G (2012) Waterborne exposure to microcystin-LR alters thyroid hormone levels and gene transcription in the hypothalamic–pituitary–thyroid axis in zebrafish larvae. Chemosphere 87:1301–1307

Yang S, Liu S, Ren Z, Jiao X, Qin S (2014) Induction of oxidative stress and related transcriptional effects of perfluorononanoic acid using an *in vivo* assessment. Comp Biochem Physiol C Toxicol Pharmacol 160:60–65

Yuan L, Lv B, Zha J, Wang W, Wang Z (2014) Basal and benzo[a]pyrene-induced expression profile of phase I and II enzymes and ABC transporter mRNA in the early life stage of Chinese rare minnows (*Gobiocypris rarus*). Ecotoxicol Environ Saf 106:86–94

Zanette J, Jenny MJ, Goldstone JV, Woodin BR, Watka LA, Bainy ACD, Stegeman JJ (2009) New cytochrome P450 1B1, 1C2 and 1D1 genes in the killifish *Fundulus heteroclitus*: basal expression and response of five killifish CYP1s to the AHR agonist PCB126. Aquat Toxicol 93:234–243

Zhang X, Gao L, Yang K, Tian H, Wang W, Ru S (2013) Monocrotophos pesticide modulates the expression of sexual differentiation genes and causes phenotypic feminization in zebrafish (*Danio rerio*). Comp Biochem Physiol C Toxicol Pharmacol 157:33–40

# Review of the Ecotoxicological Properties of the Methylenedianiline Substances

T. Schupp, H. Allmendinger, B.T.A. Bossuyt, B. Hidding, B. Tury, and R.J. West

## Contents




T. Schupp (✉)
Faculty of Chemical Engineering, Muenster University of Applied Science, Stegerwaldstrasse 39, 48565 Steinfurt, Germany
e-mail: thomas.schupp@fh-muenster.de

H. Allmendinger
Currenta GmbH and Co. OHG, Q18 105 Leverkusen, Germany
e-mail: hans.allmendinger@currenta.de

B.T.A. Bossuyt
Huntsman Europe, Everslaan 45, B-3078 Everberg, Belgium
e-mail: bart.bossuyt@huntsman.com

B. Hidding
BASF SE, GB/TC-Z570, 67056 Ludwigshafen, Germany
e-mail: bjoern.hidding@basf.com

B. Tury
Global Isocyanates Limited, Manchester, UK
e-mail: bernard@iiiglobal.net

R.J. West
The Dow Chemical Company, Midland, MI, USA
e-mail: RJWest@dow.com





P. de Voogt (ed.), *Reviews of Environmental Contamination and Toxicology, Volume 241*, Reviews of Environmental Contamination and Toxicology 241, DOI 10.1007/398_2016_6

# 1 Introduction

The methylenedianiline substances (MDAS) are a family of high production volume chemicals with annual world production volume estimated to exceed four million metric tons per year (Carvajal-Diaz 2015). More than 98 % of this production is consumed as an intermediate for the production of the methlylenediphenyl diisocyanate (MDI) substances, which are important monomers for the versatile thermoset polymer group of polyurethanes. Minor amounts of the MDAS are also consumed in manufacture of high performance polyimide fibers, and in manufacture of other specialty chemicals and resins.

The industrial-scale production of the MDAS occurs via a condensation reaction of aniline and formaldehyde, as shown in Fig. 1.

The bulk product of this reaction is commonly referred to as polymeric methylenedianiline (pMDA), and the relative proportions of the illustrated components can be controlled by adjusting the ratio of aniline and formaldehyde reactants. It should be noted here that the name "polymeric MDA" does not necessarily indicate that this reaction mixture meets the current OECD definition of "polymer". In many instances, due to > 50 % of composition coming from 4,4′-methylenedianiline, the "polymeric MDA" reaction product would not meet this OECD definition (http://www.oecd.org/env/ehs/oecddefinitionofpolymer.htm). The pMDA reaction products can be further isolated or purified by fractional distillation, making possible any number and combination of the substances listed in Table 1 (Alport et al. 2003). The 2:1 condensation products are sometimes called "2-ring-MDA" but are hereafter referred to as methylenedianiline (MDA). When isolated from the bulk pMDA reaction mixture the 2-ring MDA will typically occur

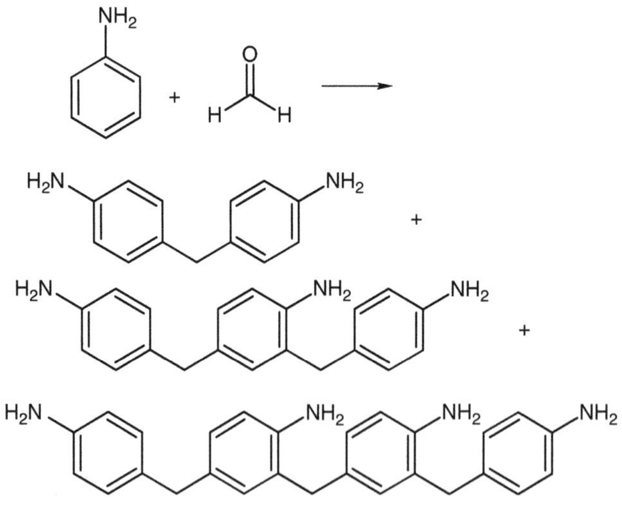


**Fig. 1** Generalized commercial synthesis route of the methylenedianiline substances (MDAS)

**Table 1** Identity of the commercially-relevant methylenedianiline substances (MDAS)

| Substance | Chemical name | Chemical Abstracts Registry Number | Synonyms | Typical % in MDA | Typical % in pMDA |
|---|---|---|---|---|---|
| Polymeric Methylene-dianiline (pMDA) | Formaldehyde, polymer with benzenamine | 25214-70-4 | pMDA | N/A | 100 |
| Methylenedianilines (MDA) | Benzenamine, 4,4′-methylenebis- | 101-77-9 | 4,4′-MDA | 90–95 | 50 |
| | Benzeneamine, 2, 4′-methylenebis- | 1208-52-2 | 2,4′-MDA | 2–5 | |
| | Benzeneamine, 2,2′-methylenebis- | 6582-52-1 | 2,2′-MDA | <1 | |
| Oligomeric methylenedianilines | 3-ring | N/A | oMDA | N/A | 25 |
| | 4-ring | | | N/A | 12 |
| | 5-ring | | | N/A | 6 |
| | Higher oligomers | | | N/A | <6 |

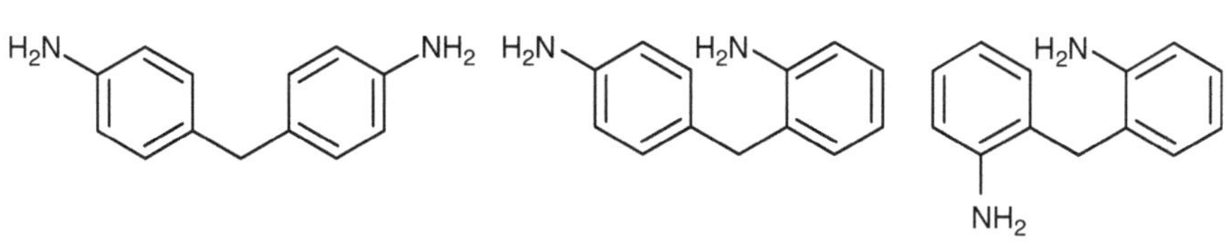


**Fig. 2** Molecular structures of the 4,4′-; 2,4′-; and 2,2′-isomers of methylenedianiline

as a mixture of three positional isomers (Fig. 2), where the 4,4′-MDA is the predominant isomer representing more than 90–95 % of the isomer mixture, with the 2,4′-MDA and 2,2′-MDA making up 2–5 % and less than 1 % of the mixture, respectively. The higher oligomer components of pMDA are sometimes named "3-ring-MDA", "4-ring-MDA" and so on, and are hereafter referred to as oligomeric MDA or oMDA.

In the evaluation of physical-chemical and toxicological properties of chemical substances, it is desirable to conduct testing on commercially-relevant substances which occur at a high-purity or as a single-component For this reason, and because it is the most commercially prominent of the methylenedianiline substances, the 4,4′-MDA substance has been intensely investigated for its physical-chemical and toxicological properties. While some properties have been studied for representative pMDA mixtures, very few studies have been conducted on the less prominent isomers of MDA and apparently none on the isolated oMDA homologues. Thus, properties of these latter substances are often inferred (i.e., by read-across) from those of 4,4′-MDA. Though mammalian toxicological hazards are not a subject of this review, it should be mentioned that 4,4′-MDA is classified by several authorities as a genotoxic carcinogen. In 2001, the European Union issued the EU risk assessment report on 4,4′-MDA which provided a summary of all physical-chemical and hazard property data known by then (European Union 2001). Since that time, and in preparation for the registration under commission regulation 1907/2006 in the EU, some more data on 4,4′-MDA were generated. Robust summaries of these past and more recent studies are now available to the public via the European Chemicals Agency (ECHA) web site (http://echa.europa.eu/web/guest/information-on-chemicals/registered-substances). In addition, the government of Canada has recently completed a draft screening level risk assessment for a grouping of the MDA, pMDA, and MDI substances (Government of Canada 2014). These and other regulatory assessments can provide good overviews of the available physical-chemical, health, and environmental properties of the MDAS family, as well as their assessed potential exposures and risks. This current review provides a deeper focus and summary of the known ecotoxicological properties of the MDAS family as available from published and private company studies available through year 2014. A similar in-depth review and summary of environmental fate properties for the MDAS family, including the aspect of bioaccumulation, will be reviewed in a subsequent review article of this journal.

The majority of studies cited in this review and in the aforementioned regulatory assessments of the MDAS have been commissioned by the International Isocyanates Institute, Inc. (III; http://diisocyanates.org/) and its private company members, and have not been published previously. For such cited references which are not from publically available sources, the ECHA web site (ECHA 2014a, b) provides robust study summaries for these studies.

## 2 Physical-Chemical Properties

The knowledge of the physical-chemical properties of the MDAS and their environmental behavior is essential to the complete understanding of the ecotoxicological effects of this substance family. The physical-chemical properties of 4,4′-MDA and pMDA are summarized in Table 2. These data for 4,4′-MDA are summarized in the EU risk assessment report (European Union 2001; McNabb JI, 1999) and in the more recent robust study summaries provided by ECHA (2014a). The substance tested consisted of 97.39 % 4,4′-MDA, 1.98 % 2,4′-MDA, 50 ppm 2,2′-MDA, and less than 10 ppm aniline.

Physical-chemical properties for pMDA are not that uniform and depend on the composition, as pMDA is an oligomeric mixture which varies slightly in composition based on the ratio of its reactants (Table 1). Robust study summaries of these pMDA properties are also provided by ECHA (2014b).

As weak bases, the water solubilities of the MDAS depend on the pH. Though the individual molecules possess two or more primary amino-groups, their pKa

**Table 2** Summary of relevant physical-chemical properties of the 4,4′-methylenedianiline (MDA) and polymeric methylenedianiline (pMDA) substances

| Property | 4,4′-Methylenedianiline (MDA) | References | Polymeric Methylenedianiline (pMDA)[a] | References |
|---|---|---|---|---|
| Physical state at 20 °C | White solid | ECHA (2014a) | Viscous liquid | ECHA (2014b) |
| Melting point (°C) | 90–92 | ECHA (2014a) | Glass transition at 0.8 and −2.7 | ECHA (2014b) |
| Boiling point (°C) | 398 ± 5 @ 101.3 kPa | ECHA (2014a) | 410.6 @101.3 kPa | ECHA (2014b) |
| Bulk density ($g/cm^3$, 20 °C) | 1.150 | ECHA (2014a) | 1.150 | ECHA (2014b) |
| Vapor pressure (Pascal) | $2.5 \times 10^{-4}$ (25 °C) | ECHA (2014a) | $<1 \times 10^{-4}$ (20 °C)<br>$1.6 \times 10^{-4}$ (50 °C) | ECHA (2014b) |
| Water solubility (g/L at 25 °C) | pH 5.3 = 2.2<br>pH 7 = 1.01<br>pH 9 = 0.84 | ECHA (2014a) | pH 7 = 0.36–1.22 | ECHA (2014b) |
| Octanol-water partition coefficient (Log Pow) | 1.55 | ECHA (2014a) | 1.2–2.7 | ECHA (2014b) |
| Dissociation constant (pKa at 20 °C) | 4.96 | ECHA (2014a) | Not determined | N/A |

[a]pMDA sample containing 58 % 2-ring-, 23 % 3- ring-, 10 % 4-ring- and 3 % 5-ring-isomers; 6 % higher oligomers

values are essentially indistinguishable by the spectrophotometric method of OECD Guideline 112 (OECD 1981). The apparent equivalence of the pKa values for these amino groups is owed to the fact that delocalization of the positive charge of the corresponding ammonium ions across the molecule is restricted by the methylene group which bridges the aromatic rings.

# 3 Environmental Behavior

The environmental fate behaviors of the MDAS are complex, and while summarized in a separate review, is it relevant to mention here the aspects which influence stability and bio-availability associated with ecotoxicological study. Briefly, like other primary aromatic amines, the MDAS react with natural organic matter in surface waters, sediment, and soil, forming covalent nitrogen-carbon bonds. The 1,4-addition of primary aromatic amines to α,β-unsaturated carbonyl compounds such as quinones is known to serve as a sink for sequestration of these amines in the environment (Parris 1980; Weber et al. 1996; Colón et al. 2002). A generalization of this reaction is illustrated in Fig. 3. Ferulic acid, a derivative of cinnamic acid which for example represents about 1 % wt of cereal plant dry mass, is another

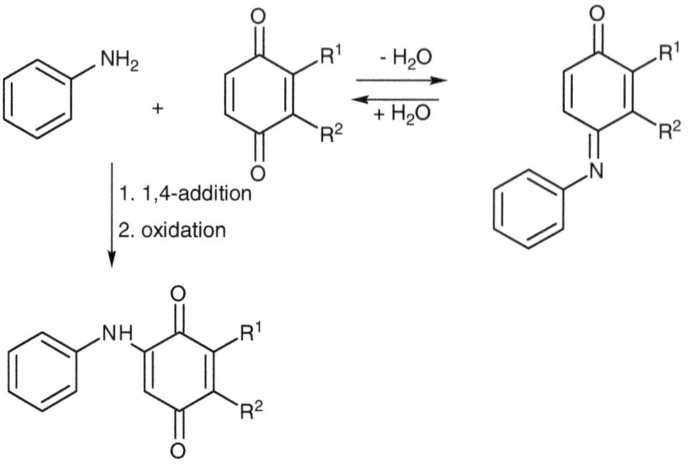


**Fig. 3** Representative reaction of primary aromatic amines with quinones in humic matter, where $R^1$ and $R^2$ indicate substituent groups of the macromolecular organic matter

reaction partner for covalent binding of primary aromatic amines in the environment, as demonstrated by Tatsumi et al. (1994).

In soil degradation tests, 4,4′-MDA showed an organic carbon-normalized adsorption coefficient ($K_{OC}$) of 3800 and 5680 in anaerobic and aerobic soils, respectively, after an equilibration time of 8 h (Cowen et al. 1998). The 8 h value achieved approximately 90 % of the value measured after 7 days equilibration time. After 7 days, part of the MDA could be desorbed when equilibrated with fresh 0.1 M $CaCl_2$ for 24 h. The high affinity of MDA to soil is most likely attributable to reversible and irreversible binding of the aromatic amino group to organic matter (Tatsumi et al. 1994; Parris 1980; Weber et al. 1996; Li et al. 2000; Colón et al. 2002).

From these examples of primary aromatic amine reactivity in environmental media, it is obvious that the availability of free MDAS has to be checked in ecotoxicological experiments to be able to draw correct and robust conclusions associated with their exposure with organisms. Vice versa, for environmental risk assessment, the potential biological (non)-availability of MDAS in eco-systems needs to be considered when data from controlled, somewhat artificial laboratory experiments are taken as benchmarks.

For the performance of ecotoxicological tests, MDAS usually need to be dissolved in solvents. Although reactions may be possible, for example, Schiff base formation, MDA was shown to be sufficiently stable in acetone, ethyl acetate, acetonitrile and diluted phosphoric acid (Cowen et al. 1996).

# 4 Ecotoxicological Properties

The ecotoxicological properties of MDA will be summarized in the following chapters, with each chapter allocated to the aquatic, benthic, and terrestrial environmental compartments and their associated standardized tests/species.

## *4.1 Toxicity to Aquatic Organisms*

In the following, the aquatic taxa bacteria, algae, daphnia and fish will be addressed. Data are summarized in Table 3.

### 4.1.1 Tests with Bacteria

Fujiwara (1981) investigated the effect of MDA on *Escherichia coli* at 25 °C. In physiological saline, 100 mg/L MDA caused a measurable reduction in cell count over 5 days, whereas 50 mg/L was the no-observed-effect concentration (NOEC). In nutrient broth, 100 mg/L MDA did not cause a reduction in cell count compared to the control.

Caspers et al. (1986) investigated the inhibition of activated sludge by 4,4′-MDA. The composition of the test substance was 99.7 % 4,4′-MDA, 0.25 % 2,4′-MDA and 0.05 % 2,2′-MDA. The test was performed according to guideline OECD No. 209 (1984). After a 3 h contact time, oxygen consumption was measured for 0.5 h. Test concentrations were 1, 10 and 100 mg/L (nominal). At each concentration tested, the inhibition was 15 %. The MDA concentration was below the limit of solubility, and the lack of a dose–response is somewhat odd, but the validity of the test system was checked with the positive control 3,5-Dichlorophenol. The authors concluded the 3 h-$EC_{50} > 100$ mg/L. It should however be noted that $EC_{50}$ values for bacteria are dependent on the biomass loading of the tested inoculum. Evidence for inhibition of bacterial respiration has been observed in biodegradation screening tests employing as little as 10 mg/L 4,4′-MDA and 30 mg/L (dry solids) activated sludge (unpublished data of The Dow Chemical Company).

Bringmann and Meinck (1964) reported that MDA inhibits the digestion of glucose by *Pseudomonas putida* at a concentration of about 15 mg/L. The digestion of glucose resulted in a decrease of the pH value from pH = 7.5 to pH = 6.0. The lowest test concentration, which resulted in a higher pH value at the end of the test compared to the control, was called inhibition concentration. Details on the composition of the test substance are not provided.

Kaiser and Palabrica (1991) investigated the effect of 4,4′-MDA (99 % pure) on the luminescent bacterium *Photobacterium phosphoreum* (i.e., *Vibrio fischeri*). The

**Table 3** Summary of 4,4′-methylenedianiline (MDA) toxicity to aquatic organisms

| Species | Endpoint and duration | Result (mg/L) | References |
|---|---|---|---|
| Bacteria *Escherichia coli* | 5 days, cell count | LOEC = 100<br>NOEC = 50 | Fujiwara (1981) |
| Bacteria activated sludge | 3 h, respiration inhibition, | $EC_{15} = 1$<br>$EC_{50} > 100$ | Caspers et al. (1986) |
| Bacteria *Pseudomonas putida* | 16 h, growth | $EC_{20} = 15$[a] | Bringmann and Meinck (1964) |
| Marine bacteria *Vibrio fischeri* | 0.5 h, luminescence | $EC_{50} = 6.6$ | Kaiser and Palabrica (1991) |
| Bacteria *Ochrobactrum anthropi* | 35 days, respiration inhibition | $EC_{50} = 53$ | Kim et al. (2002) |
| Fungus *Aspergillus* sp. | 35 days, respiration inhibition | $EC_{50} = 53$ | Kim et al. (2002) |
| Green algae *P. subcapitata* | 72 h | $E_bC_{50} = 5.34$<br>$E_rC_{50} = 14.4$<br>$NOEC_b = 0.93$<br>$NOEC_r = 9.3$ | Mitsubishi Chemical Safety Institute (2008a) |
| Green algae *S. subspicatus* | 96 h; growth | $EC_{20} = 31$[a] | Bringmann and Meinck (1964) |
| Green algae *S. subspicatus* | 72 h; $E_rC_{50}$ | 21 | Rufli and Mueller (1985a) |
| Green algae *S. subspicatus* | 72 h; | $E_bC_{50} = 9.8$<br>$E_rC_{50} = 11.0$<br>$E_bC_{10} = 2.4$<br>$E_rC_{10} = 0.3$ | European Union (2001) |
| Green algae *C. pyrenoidosa* | N.L.[b] | 1–10 | Rhône-Poulenc Chimie (1977) |
| Cyanobacterium *S. cedrorum* | N.L.[b] | 1–10 | Rhône-Poulenc Chimie (1977) |
| Marine Diatom *N. fustulum* | N.L.[b] | 10–100 | Rhône-Poulenc Chimie (1977) |
| Crustacean *Daphnia magna* | $EC_{50}$ | 5.7 (24 h) | Rufli and Mueller (1985c) |
| Marine Crustacean *Moina macrocopa* | $EC_{50}$ | 2.3 (24 h) | Fujiwara (1982) |
| Marine Crustacean *Moina macrocopa* | 14-days | 0.15 | Fujiwara (1982) |
| Crustacean *Daphnia magna* | $EC_{50}$ | 0.25 (48 h) | Bringmann and Meinck (1964) |
| Crustacean *Daphnia magna* | $EC_{50}$ | 8.08 (24 h);<br>2.47 (48 h) | Mitsubishi Chemical Safety Institute (2008b) |
| Crustacean *Daphnia magna* | $EC_{50}$ | 0.19–0.6 (24 h);<br>0.019–0.06 (48 h) | Mitsubishi Chemical Safety Institute (2008c) |
| Crustacean *Daphnia magna* | 21-day | NOEC = 0.00525<br>LOEC = 0.0182 | Mitsubishi Chemical Safety Institute (2008c) |

(continued)

**Table 3** (continued)

| Species | Endpoint and duration | Result (mg/L) | References |
|---|---|---|---|
| Crustacean *Daphnia magna* | $EC_{50}$ | 1.5 (24 h); 0.35 (48 h) | Salinas (2011) |
| Crustacean *Daphnia magna* | $EC_{50}$ | 2.19 (24 h); 0.46 (48 h) | Salinas (2012)[c] |
| Mollusca *Limnea stagnalis* | 20-h $LC_{50}$ | 210 | Rhône-Poulenc Chimie (1977) |
| Zebrafish *Danio rerio* | 96-h $LC_{50}$ | 42 | Rufli and Mueller (1985b) |
| Rainbow trout *Oncorhynchus mykiss* | 96-h $LC_{50}$ | 39 | Rufli and Mueller (1985c) |
| Zebrafish *Danio rerio* | 96-h $LC_{50}$ | 65 | Caspers et al. (1986) |
| Golden orfe *Leuciscus idus* | 96-h $LC_{50}$ | 53 | Munk and Kirsch (1988) |
| Medaka *Oryzias latipes* | 96-h $LC_{50}$ | 20.6 | Mitsubishi Chemical Safety Institute (2008d) |

[a]Assumed to be an $EC_{20}$ (see text)
[b]Noxious limit (see text)
[c]56.9 % 4,4′-MDA, 7.1 % 2,4′-MDA, 26 % 3-ring MDA, 10 % 4-ring MDA and higher oligomers

$EC_{50}$, determined as the concentration resulting in 50 % reduction in light emission compared to control after 30 min contact time, was 6.6 mg/L.

In the scope of biodegradation tests with 4,4′-MDA, Kim et al. (2002) demonstrated that at a concentration of 30 mg/L MDA, $CO_2$ evolution reached about 70–80 % of the theoretical yield over 35 days with non-adapted activated sludge as well as with enrichment cultures of *Ochrobactrum anthropi* or *Aspergillus* sp. These enrichment cultures were generated by the incubation of soil extracts taken from areas where dehydrated activated sludge of the waste water treatment plant was buried with MDA as the sole carbon source. However, at concentrations of 50, 100 or 300 mg/L MDA, $CO_2$ evolution reached levels of about 50, 20 and 5 % after 35 days, respectively. Data were presented in diagrams and, therefore, any further interpretation has to be regarded with care. If it is assumed that 30 mg/L MDA had no effect on the microorganisms, dosages of 50, 100 and 300 mg/L reduced the $CO_2$ evolution down to levels of 71, 21 and 7 % for *O. anthropi* and to 57, 14 and 7 % for *Aspergillus* sp. A rough logit analysis of the graphical data given in the paper results in $EC_{50}$ values of about 53 mg/L for *O. anthropi* and *Aspergillus* sp., however the concentration of biomass associated with these tests was not quantitatively determined. From the dose–response curves it is not clear whether or not 50 mg/L represent already an inhibition of the microorganisms or whether there is simply a substrate overload; 100 mg/L, however, is clearly inhibiting for *Aspergillus* sp., and for *O. anthropi* the formation of $CO_2$ comes to a halt after

25 days. Further, at 30 mg/L the $CO_2$ production rate is virtually the same between the non-adapted sludge and the enrichment cultures; this result indicates toxicity of MDA to MDA-digesting microorganisms.

### 4.1.2 Tests with Algae

The toxicity of 4,4′-MDA to a freshwater single-celled algae (*Pseudokirchneriella subcapitata*; previously *Selenastrum capricornutum*) was investigated by the Mitsubishi Chemical Safety Institute in 2002, and an English translation was made available as International Isocyanate Institute report 11545 (Mitsubishi Chemical Safety Institute 2008a). The test was performed according to OECD Guideline No. 201 (2006). The purity of the test substance was 99.6 %, and test concentrations were checked by HPLC-UV. The test substance concentration was constant (±20 %) over the exposure time. In the European Union, the median growth rate inhibition concentration ($E_rC_{50}$) is the preferred endpoint; in this article, the endpoint of median biomass inhibition concentration ($E_bC_{50}$) is provided additionally. Against biomass, after 72 h the $E_bC_{50}$ was 5.34 mg/L (95 % C.I. 3.57–7.98 mg/L) and the $NOEC_b$ was 0.93 mg/L. In terms growth rate, the $E_rC_{50}$ was 14.4 mg/L (24–72 h) and the $NOEC_r$ was 9.3 mg/L.

In a study with the colonial freshwater alga, *Scenedesmus subspicatus*, the 72 h $E_rC_{50}$ was 21 mg/L (95 % C.I. 11–29 mg/L) for 4,4′-MDA (Rufli and Mueller 1985a). The test was performed according to OECD Guideline 201 (1981). The test substance was named TK 10504 (commercial grade) and claimed to be 4,4′-MDA. Data on purity are not provided. However, the European Union cites this report (2001) and mentions a purity of 95.5–98 %. Results were based on measured concentrations and the data were provided; the analytical method is not mentioned.

Caspers and Mueller (1992) investigated the toxicity of pMDA on *Scenedesmus subspicatus*. After 72 h exposures, the $E_bC_{10}$ and $E_bC_{50}$ values were 2.4 and 9.8 mg/L, and the $E_rC_{10}$ and $E_rC_{50}$ values were 0.3 and 11.0 mg/L, respectively.

Bringmann and Meinck (1964) reported a limit concentration of 31 mg/L for the inhibition of growth for *Scenedesmus* for 4 days exposure. It is assumed that this level is equal to an $EC_{20}$.

A pMDA sample containing 60 % wt 4,4′-MDA was tested against various algae/cyanobacteria (Rhône-Poulenc Chimie 1977). In those tests, a "noxious limit" was established, which was a concentration were the growth of the organisms was inhibited against a control. Tests were stated to be "run for a time sufficient to detect multiplication of organisms in the control". For the green alga *Chlorella pyrenoidosa* and the blue-green alga *Synechocystis cedrorum* (i.e., *Cyanobacterium cedrorum*) the noxious limit was in the range from 1.0 to 10.0 mg/L. For the marine diatom *Nietzschia frustulum* the noxious limit fell between 10.0 and 100.0 mg/L. Test concentrations were verified to be controlled within initial nominal values using a photometric method.

As the Mitsubishi studies are the only ones conducted and documented under Good Laboratory Practice (GLP) standards for quality assurance (Mitsubishi Chemical Safety Institute 2008a) these data could be regarded as the most reliable for environmental risk assessment. The data generated by Rhône-Poulenc Chimie (1977) add value in so far as there seems not to be a significant greater sensitivity of cyanobacteria and diatoms compared to green algae.

### 4.1.3 Tests with Crustacea

With the freshwater crustacean *Daphnia magna*, data from several acute and chronic test exposures are available. Rufli and Mueller investigated the acute toxicity of 4,4′-MDA against *Daphnia magna* (Rufli and Mueller 1985b) according to OECD Guideline 202 (1981). The exposure time was 24 h, whereas the current guideline requires a 48 h exposure. Due to a resulting odd dose–response behavior (Fig. 4), an $EC_{50}$ could not be established. The authors reported 24 h $EC_0$ and $EC_{100}$ values of $< 3.2$ and $> 100$ mg/L, respectively. However, taking the three lowest concentrations only, a more meaningful dose–response relationship becomes evident, and the 10 mg/L exposure would appear to cause about 75 % immobility. Exposure concentrations of this study are based on measured concentrations, but the analytical method was not mentioned in that report. By probit analysis, an $EC_{50}$ of 5.7 mg/L can be calculated; however, using this reduced data set a 95 % C.I. cannot be derived.

Fujiwara (1982) reported a 24 h $EC_{50}$ of 2.3 mg/L for *Moina macrocopa*, a common aquatic invertebrate in the Asia-pacific region which tolerates high salinity (95 %-C.I.: 1.8–3.0 mg/L). The test substance was claimed to be extra-pure with no

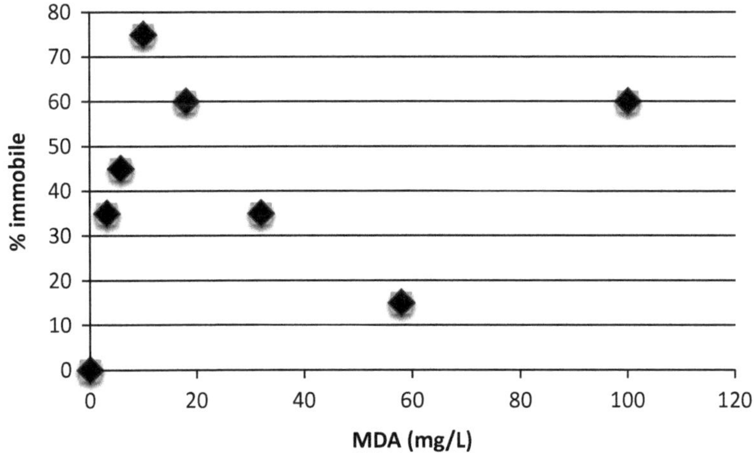


**Fig. 4** Immobility responses associated with acute (24 h) exposures of 4,4′-MDA to *Daphnia magna* (Rufli and Mueller 1985b)

further details provided. The composition and pH of the test water were described. For each exposure concentration, 4 replicates of 5 animals each were prepared. The reference toxicant potassium dichromate ($K_2Cr_2O_7$) served as positive control. There is no information concerning an analytical method. Therefore, results reported are most likely based on nominal concentrations.

In an OECD Guideline 202 study conducted in compliance with GLP, the acute $EC_{50}$ for *Daphnia magna* was 2.47 mg/L after an exposure time of 48 h (Mitsubishi Chemical Safety Institute 2008b). After 24 h, the test solutions were renewed (semi-static). The 24 h $EC_{50}$ was 8.08 mg/L (95 % C. I.: 5.23–12.8 mg/L). After an exposure time of 48 h, the highest concentration at which no effects were observed was 0.2 mg/L, and the $EC_{100}$ was 200 mg/L. The 95 % C.I. for the 48 h $EC_{50}$ was 1.27–4.4 mg/L. Concentrations of the test substance were verified analytically with HPLC-UV. The purity of the test-substance was 99.6 %. Another more recent OECD Guideline 202 (OECD 2004) study on the acute toxicity of 4,4′-MDA to *Daphnia magna* was issued in 2011 (Salinas 2011). The 48 h $EC_{50}$ was determined as 0.35 mg/L (95 % C. I. 0.18–0.71 mg/L), whereas the 48 h $EC_0$ and 48 h $EC_{100}$ were 0.22 and 2.2 mg/L, respectively. The purity of the test substance was 98 %. According to HPLC-MS analyses, impurities consisted of an isomer to 4,4′-MDA, as well as some 3-ring MDA and N-methylated by-products. The concentrations of the test solutions maintained over the exposures were verified with HPLC-UV. This test was performed under static conditions.

The acute toxicity of pMDA to *Daphnia magna* was checked in an OECD Guideline 202 test (Salinas 2012). The test substance was characterized by IR, UV, HPLC-MS, GC and $^1$H-NMR. The tested product consisted of 55.9 % 4,4′-MDA, 7.1 % 2,4′-MDA, 0.3 % 2,2′-MDA, 26.1 % 3-ring MDA and 9.9 % 4-ring MDA and 0.7 % impurities, from which N-methyl-methylene-4,4′-dianiline and N-formyl-methylene-4,4′-dianiline could be identified. To avoid potential for physical toxicity associated with undissolved particulate matter in test solutions, pMDA was dissolved in peroxide-free THF ($H_2O_2 \leq 1$ ppm) and the THF was evaporated in a rotating round bottom flask. M4 medium was added to the dried flask and slowly shaken (50 rpm) for 24 h in the dark. Finally, the resulting test solutions were filtered through a 0.2 μm membrane. Analyses were performed to determine total dissolved organic carbon (DOC) and 4,4′-MDA concentrations by HPLC-UV. The derived 48 h $EC_{50}$ of 0.46 mg/L (nominal) or 0.26 mg/L (as measured 4,4′-MDA), indicate that the acute toxicity of pMDA to *D. magna* is in the same order of magnitude as that for pure 4,4′-MDA.

In addition to the aforementioned acute tests with aquatic crustacea, two chronic exposure tests have been conducted to determine potential effects on survival and reproduction. Fujiwara (1982) performed a 14 days chronic test with *Moina macrocopa*. The test period covered three generations of the organism. Animals were not older than 24 h at the beginning of the test. Per test concentration, four replicates with ten individuals each were prepared. The test solutions were renewed every 48 h, and the organisms were fed with unicellular green algae daily. The 14 days NOEC for reproduction was 0.15 mg/L. There is no information concerning an analytical method; therefore, results reported are most likely based on nominal exposure concentrations.

In 2002, Mitsubishi Laboratories performed a chronic study (21 day) with *Daphnia magna* according to GLP and OECD Guideline 211 (1998); the data were made available in English in 2008 by the International Isocyanate Institute (Mitsubishi Chemical Safety Institute 2008c). Test solutions were renewed daily. The purity of the test substance was 99.6 %, and exposure concentrations were verified using HPLC-UV analyses. In the scope of that test it was observed that the 48 h $EC_{50}$ was between 0.019 and 0.06 mg/L; the authors reported a 21 day NOEC of 0.00525 mg/L and a lowest observed effect concentration (LOEC) of 0.0182 mg/L based on the reproduction endpoint.

Concerning acute toxicity and LC/EC50 values for crustacea, *D. magna* is the most sensitive of all species tested with the MDAS, owing to the 48 h $EC_{50}$ of 0.019–0.06 mg/L (Mitsubishi Chemical Safety Institute 2008c) and the results of the Salinas (2011) study, yielding a 48 h-EC50 of 0.35 mg/L. This latter value is not significantly different to that of pMDA (0.46 mg/L), tested in the same laboratory. The highest 48 $EC_{50}$ value reported for daphnia is 2.47 mg/L (Mitsubishi Chemical Safety Institute 2008b). Therefore, the MDA acute toxicity to crustacea spreads over two orders of magnitude if results 48h data from acute and chronic tests are considered; focused on data from acute tests only, the difference between 48 h $EC_{50}$ values is about 10. Shortcomings in the tests performed were not identified, and impurities are not a likely explanation as this bulk industrial chemical commodity is synthesized via a well known, standard route. The 24 h $EC_{50}$ values for $K_2Cr_2O_7$ were in the range required by OECD Guideline 202. In the Mitsubishi Chemical Safety Institute acute and chronic tests (2008b, c), the water was renewed daily. The only remaining difference identified is the fact that in the chronic test, the daphnids were fed daily with *Chlorella vulgaris*. It remains to be found out whether this difference to the acute test is at least partially responsible for the differences in results.

### 4.1.4 Test with Vertebrates (Fish)

Rufli and Mueller reported the acute toxicity of MDA against zebrafish (*Danio rerio*) (1985c). The static test was performed according to OECD Guideline 203 (1984). This study resulted in a 96 h $LC_{50}$ of 42 mg/L (95 % C.I.: 35–51 mg/L). Results were based on measured concentrations; however neither the analytical method for concentrations verification nor the accurate purity of the test substance was given. Rufli and Mueller (1985d) also reported the acute toxicity of 4,4′-MDA against rainbow trout (*Oncorhynchus mykiss*; *formerly Salmo gairdneri*), following OECD Guideline 203 (1984) under a static exposure procedure. The 96 h $LC_{50}$ was 39 mg/L (95 % C.I.:20–134 mg/L). Results are based on measured concentrations, but again neither the analytical method for concentrations verification nor the accurate purity of the test substance was given.

A further report on the toxicity of MDA against zebra fish is reported by Caspers et al. (1986), following OECD Guideline 203 (1984). The authors report a 96 h $LC_{50}$ of 65 mg/L, and 96 h $LC_0$ and $LC_{100}$ values of 40 and 70 mg/L, respectively. The test substance consisted of 99.7 % 4,4′-MDA, 0.25 % 2,4′-MDA and 0.05 % 2,2′-MDA. Maintenance of the exposure concentrations over 96 h was verified using dissolved organic carbon (DOC) analyses.

Munk and Kirsch (1988) reported the acute toxicity of MDA against golden orfe (*Leuciscus idus*). The test procedure followed the German standard DIN 38412 (DIN 1980). Chloro acetamide was used as a positive control. The dissolved oxygen content of the water was throughout > 60 % of the air saturation value, pH was maintained at 8, and the water temperature was held at 20 ± 1 °C. Water hardness and salinity were kept in the required range. Purity of the test substance was claimed to be at least 96 %. The 96 $hLC_{50}$ was determined by graphical interpolation to be 53 mg/L. The 96 h $LC_0$ was 21.5 mg/L and the 96 h $LC_{100}$ was 100 mg/L, where derivation of these endpoints is based on nominal concentrations. At higher concentrations, the test substance was noted to have precipitated from solution.

In an English-translated summary report of original studies of the Chemicals Inspection and Testing Institute-Japan, a 48 h $LC_{50}$ of 32 mg/L was reported for the orange-red killifish, *Oryzias latipes* (CITI 1992). The testing regime was semi-static, with renewal of the exposure solutions every 8–16 h, and followed the Japanese standard JIS K 0102-1986-71. The Mitsubishi Chemical Safety Institute similarly investigated the acute toxicity of MDA with *Oryzias latipes* in 2002 according to OECD Guideline 203 (1992) and in compliance with GLP. The translated report was issued by the International Isocyanate Institute (Mitsubishi Chemical Safety Institute 2008d). Applying a semi-static method with daily renewal of the test solutions, a 96 h $LC_{50}$ of 20.6 mg/L was reported (95 % C.I.: 16.7–25.3 mg/L). These results were based on measured concentrations (HPLC-UV) and the test substance had a purity of 99.6 %.

### 4.1.5 Tests with Mollusca

The acute toxicity of a pMDA sample containing 60 % 4,4′-MDA was examined using 20 h static aquatic exposures to the Great Pond Snail (*Limnea stagnalis*). The tests were performed in reconstituted river water in the dark at 20 °C, with pH of 8.0 and dissolved oxygen maintained at ≥ 80 % air saturation. A cohort of 6–8 days post-hatch snails was investigated for egg laying performance, and a 1–4 days post-hatch cohort was checked for survival. The $LC_{50}$ for the post exposure 4 days survival of young snails was 210 mg/L, and for egg laying performance (fecundity) the $LC_{50}$ was 220 mg/L (Rhône-Poulenc Chimie 1977). Aside from this study, no other studies of MDAS with freshwater or marine mollusca are known. Based on this single study, this animal phylum would appear to be less sensitive to the MDAS substances than are the crustacea.

## 4.2 *Toxicity to Benthic Organisms*

While 4,4′-MDA was under evaluation in the European Union, effects of chemicals to benthic sediment organisms gained interest. This was particularly true for substances highly hydrophobic and/or reactive substances such as the MDAS which have high affinity for sediment and associated organic matter. In a research program of the German Umweltbundesamt (UBA; Environmental Protection Agency), the sediment blackworm *Lumbriculus variegatus* was revealed to be particularly sensitive against primary aromatic amines (Riedhammer and Schwarz-Schulz 2001). As a consequence, 4,4′-MDA was tested against the sediment organisms *Lumbriculus variegatus*, *Chironomus riparius* and Hyalella azteca, representing three different taxa of sediment organisms (oligochaete, insects and amphipods) and playing an important role in the aquatic food chain and in organic matter turnover. Data concerning benthic organism are summarized in Table 4.

### 4.2.1 Tests with an Oligochaete

The benthic oligochaete, *L. variegatus* (blackworm) inhabits a wide range of sediment types and is typically present in shallow water sediments of slowly flowing rivers which are rich in organic matter. It is feeding on mainly subsurface organic matter of dead and decaying organisms, and because primary aromatic amines are known to be covalently bound with such organic matter, the potential effects on this organism are of relevance. Egeler and Ginzburg (2001) investigated the toxicity of 4,4′-MDA to *L. variegatus*. The study was conducted in compliance with GLP, but since the now standardized OECD Guideline 225 (2007) for sediment-water *Lumbriculus* toxicity tests was still under development, and

**Table 4** Summary of toxicity of 4,4′-methylenedianiline (MDA) to sediment organisms

| Species | Duration | Endpoint and value (mg/L) | References |
|---|---|---|---|
| Oligochaete *Lumbriculus variegatus* | 28 days | NOEC = 25.2<br>LOEC = 50.3 | Egeler and Ginzburg (2001) |
| Oligochaete *Lumbriculus variegatus* | 28 days | NOEC < 3.75[a],<br>NOEC = 30[b]<br>LOEC = 60[b] | Egeler (2002) |
| Insect larvae *Chironomus riparius* | 28 days | NOEC = 500<br>LOEC = 1000 | Egeler and Gilberg (2005a) |
| Amphipod *H. azteca* | 28 days | NOEC = 41.4[c]<br>NOEC = 90.9[d]<br>LOEC = 90.9[c]<br>LOEC = 200[d] | Egeler and Gilberg (2005b) |

[a]Based on statistics (see text)
[b]Biological interpretation (see text)
[c]Survival
[d]Biomass

because *Lumbriculus* turned out to be most sensitive sediment organism against MDA with a non-monotonic dose–response, a more detailed explanation on test design and results will be provided in the following section.

In a first test with *L. variegatus*, the artificial sediment consisted of 75 % quartz sand, 20 % clay and 5 % sphagnum peat. The peat content was chosen to represent a moderate content of organic matter as indicated at that time. Further, between 0.05 and 1 % pure $CaCO_3$, 0.4 % TetraMin® fish food (Tetra Werke, Melle, Germany) and 46 % water were added per sediment dry weight. While the test was running, further TetraMin® was added on the surface of the sediment at the beginning of weeks 2, 3 and 4. Reconstituted water was used according to OECD guideline 203 (1992) Annex 2. During the test, the temperature was maintained at $20 \pm 2$ °C, and a 16/8 h light (100–1000 Lux)/dark cycle was applied. Air was slowly passed through the aqueous layer at one bubble per second. As MDA can bind covalently to sediment, spiked sediment was allowed to equilibrate for 2 days before the test organisms were added. This equilibration time was chosen to ensure that added organisms were exposed against both organic matter-bound and freely dissolve MDA. Per dose group, three replicate exposures were prepared with ten synchronized worms added to each of the 1 L test vessels, filled with equilibrated sediment and reconstituted water. Nominal concentrations of 4,4′-MDA (purity 97.9 %) were 0, 1.6, 3.1, 6.3, 12.6, 25.2, 50.3 and 100.7 mg/kg sediment dry weight. The spiking solutions, overlying water, and sediment pore-water were analyzed by HPLC-UV. During the test, the dissolved oxygen in the overlying water was at least 74 % of the air saturation value, and the pH was kept between 6.0 and 7.0. At the end of the 28 days exposure period, the number of worms and total worm dry weight were analyzed by ANOVA and Dunnett's *t*-test. In the control group, mortality did not exceed 7 %. For reproduction (total number of worms and number of regenerated worms), biomass, and number of worms with new posterior and anterior end, a clear dose–response pattern could not be observed (Figs. 5–7). Due to the reactivity of MDA with organic matter, and possibly also due to its biodegradation in the sediment-water test system, its concentration in the aqueous phase declined over

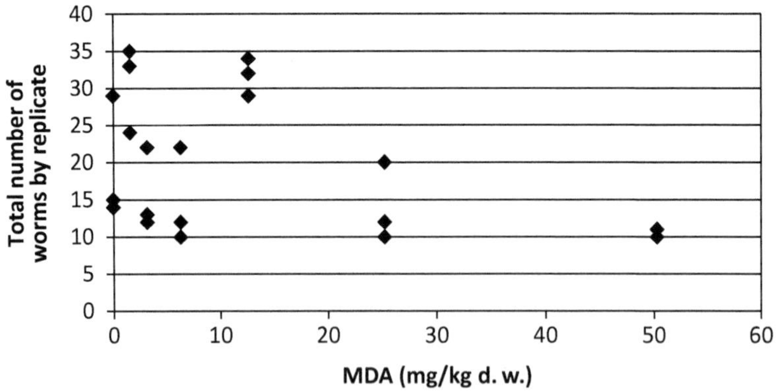


**Fig. 5** Representative plot of number of *L. variegatus* worms recovered from replicate 4,4′-methylendianiline (MDA) sediment-water exposure vessels after 28 days (adapted from Egeler and Ginzburg 2001)

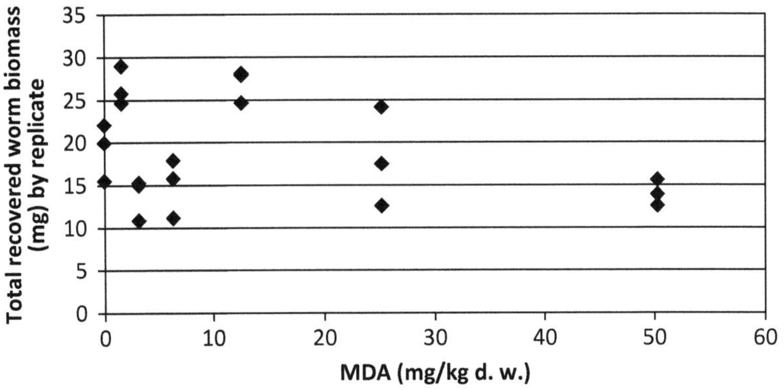


**Fig. 6** Representative plot of total worm biomass recovered from replicate 4,4′-methylendianiline (MDA) sediment-water exposure vessels after 28 days (adapted from Egeler and Ginzburg 2001)

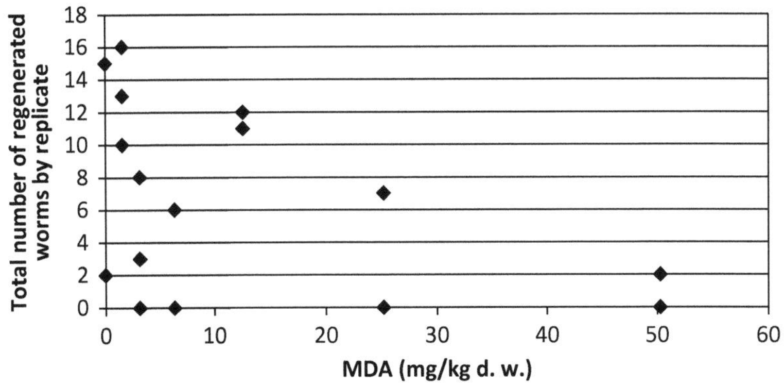


**Fig. 7** Representative plot of total of total number of *L. variegatus* worms with new posterior/ anterior ends recovered from replicate 4,4′-methylendianiline (MDA) sediment-water exposure vessels after 28 days (adapted from Egeler and Ginzburg 2001)

time. The mean recovery of MDA was about 4 % of that initially applied after 28 days. Based on nominal concentrations, the NOEC and LOEC values for biomass were 50.3 and 100.7 mg/kg d. w., respectively, whereas for reproduction the values were 25.2 and 50.3 mg/kg d. w. Based on measured concentrations after 28 days, the NOEC and LOEC were 2.1 and 4.1 mg/kg d. w. against biomass and 1.0 and 2.1 mg/kg d. w. for reproduction. For lower nominal concentrations, extractable MDA was not detectable at the end of the test period (<0.1 mg/kg d. w.). In freshly-spiked sediment, recovery of MDA was 82–102 % of the applied concentration. Therefore, the extraction and analytical methods as such were effective. The disappearance of the MDA in these test systems was most likely attributable to its irreversible binding to organic matter of the formulated sediment, and perhaps also biodegradation in both sediment and water layers. In the previously described test with *L. variegatus*, it was speculated the iterative feeding with

TetraMin® could predispose the test organisms towards selective feeding on this uncontaminated food versus the MDA-associated sediment organic matter. In such a case, the true toxicity of MDA-associated sediment organic matter may be misrepresented. A second study was performed to check whether sediment amended with an organic matter food source– here nettle powder (*Urtica* sp.)—instead of semi-continuous feeding would deliver different results (Egeler 2002). Nominal concentrations of 4,4′-MDA (purity 97.9 %) were 0, 3.75, 7.5, 15, 30 and 60 mg/kg sediment dry weight. Due to high variability among replicates observed in the previous test, six replicates were used for the control and three replicates for each exposure concentration. The test solutions used to spike the formulated sediment with MDA were verified by HPLC-UV, but further analyses of pore-water and overlying water was deemed not necessary due to the reactivity of MDA to sediment organic matter.

A dose–response curve could not be fitted to the data due to high variations between dose groups (Figs. 8 and 9). For the total recovered biomass endpoint, the lowest test concentration of 3.75 mg/kg d. w. was the LOEC after 28 days. For reproduction, the 28 days NOEC was 3.75 mg/kg and the 28 days LOEC was 7.5 mg/kg d. w. sediment. For regenerated worms with new anterior and posterior ends, the effect of MDA was more prominent. In the control vessels, the number of complete new worms ranged from 2 to 8. In all exposure vessels, there was at maximum one complete new worm.

The results from these two tests with *L. variegatus* would indicate that the feeding regimen employed can influence the derived effective concentrations for biomass, reproduction, and regeneration of this sediment-feeding worm. The test which employed feeding via the overlying water appeared to result in higher effective concentrations for the MDA substance. However, it cannot be determined how difference in feeding modes among these two studies could influence other factors such as microbial activity (biodegradability MDA) and ammonia production

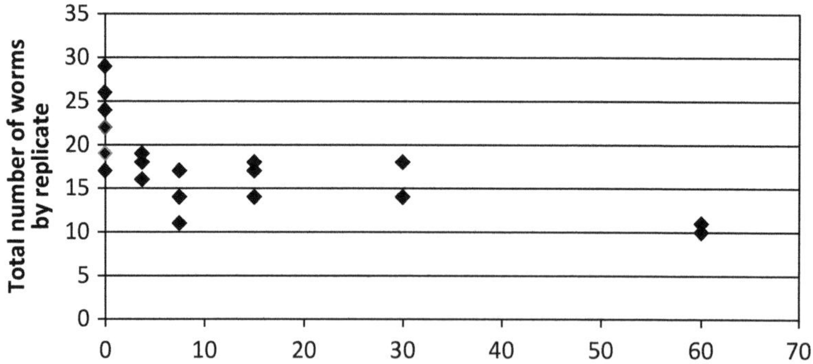


**Fig. 8** Representative plot of total of total number of *L. variegatus* worms recovered from replicate 4,4′-methylendianiline (MDA) sediment-water exposure vessels after 28 days (adapted from Egeler 2002)

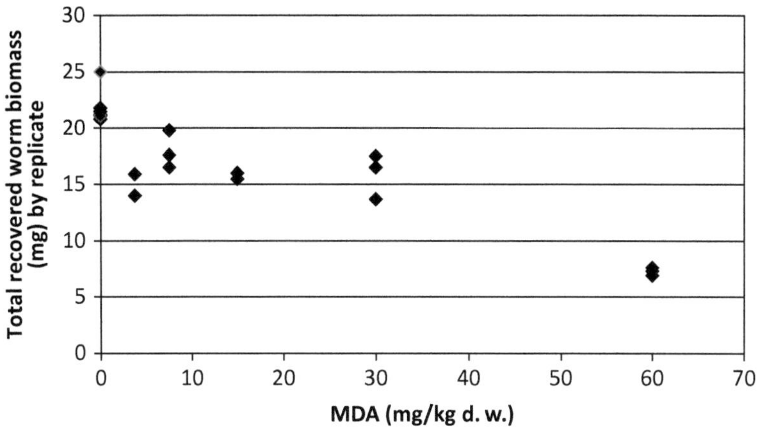


**Fig. 9** Representative plot of total worm biomass recovered from replicate 4,4′-methylendianiline (MDA) sediment-water exposure vessels after 28 days (adapted from Egeler 2002)

as confounding toxicant in these sediment-water test systems. Nevertheless, this benthic oligochaete was shown to be most sensitive to MDA in both tests, compared to other species evaluated.

### 4.2.2 Tests with an Insect Larvae

Larvae of the harlequin fly (*Chironomus riparius*) are living in and on sediments, and are feeding on organic matter deposited in the sediment during this life cycle stage. As such, they too are a relevant test organism for assessing potential hazard of substances such as the MDAS which bind with sediment organic matter. In an OECD Guideline 218 study, Egeler and Gilberg (2005a) tested the chronic toxicity of 4,4′-MDA (purity 97.9 %) to *C. riparius*. Prior to test substance addition, the sediment was spiked with 0.4–0.5 % stinging nettle powder as the sole food source. As West et al. (2004) discovered, the spiking of sediment with nettle powder can produce significant amounts of ammonia. As a result, in this study a tighter control of pH values was maintained to maximize speciation of any ammonia produced as the less toxic ammonium ion. In addition, aeration of the overlying water layer was doubled from that employed in the *L. variegatus* tests from one to two bubbles of air per second, so that any un-ionized ammonia (gas) would be more readily expelled. Larvae were added 2 days after spiking the sediment with MDA. This procedure ensured that part of the test substance is already bound to sediment, while another part is still available in the aqueous phase. Nominal test concentrations of 4,4′-MDA (purity 97.9 %) were 62.5, 125, 250, 500 and 1000 mg/kg d. w. sediment. Due to the experience with previous tests, MDA was analyzed in stock solutions only, but not in sediment and overlying water. The pH was kept in the range of 7.3–8.3 by occasional addition of diluted HCl (pH up to 9 would be acceptable according to the guideline). The ammonium content in the overlying water ranged from 0.2 to 16.1 mg/L in the exposures employed in this test. For the emergence ratio endpoint, the 28 days NOEC was 500 mg/kg dry weight, and 1000 mg/kg d. w. was the 28 days LOEC. For the development rate endpoint, 1000 mg/kg d. w. was determined to be the 28 days NOEC. Thus, the larvae of the *C. riparius* insect appear to be much less sensitive to MDA than is the oligochaete *L. variegatus*.

### 4.2.3 Test with an Amphipod

A third organism has been tested with chronic exposures to MDA, to enable an analysis of species sensitivity across three taxa of benthic organisms and their associated feeding modes. The amphipod crustacean *Hyalella azteca* shows some subsurface deposit feeding, but has a stronger affinity to feeding on organic matter deposits at the surface of sediments than the previously describe benthic organisms. The *H. azteca* organism is very sensitive to ammonia (Whiteman et al. 1996); the sensitivity depends on the pH value of the test medium. At a nominal concentration of 50 mg/L $NH_4^+$ (148 mg/L $NH_4Cl$) in the overlying water, all exposed amphipods survived a 4 days exposure period in M4 medium at pH values not higher than 8.5. When the pH value was raised to 9, all exposed organisms died within 24 h (Egeler 2004).

Egeler and Gilberg (2005b) tested the chronic toxicity of 4,4′-MDA to *H. azteca*. The testing regime was performed in compliance with GLP and followed OECD Guideline 218 (OECD 2004). The chironomus test conditions were adapted to testing with *H. azteca*, as an OECD guideline for testing with this organism does not exist. In this adapted test, the amount of stinging nettle powder incorporated into the sediment was reduced from the guideline recommended 0.5 % d.w. to 0.25 % d.w. This was thought to reduce potential for ammonia formation from this N-containing organic matter, and 0.25 % cellulose (which contains no nitrogen) was added to replace amount of eliminated nettle. These sediment-incorporated food sources were equilibrated in the test system before addition of the test substance. Nominal concentrations of 4,4′-MDA (purity 97.9 %) were 8.5, 18.8, 41.3, 90.9 and 200.0 mg/kg sediment d. w.; spiking solutions were analyzed by HPLC-UV, but no analysis was performed for the sediment and the overlying water due to the known reactivity of MDA to sediment organic matter. After a 2 day equilibration period organisms were added, and pH was controlled daily between 7.4 and 8.3 by occasional addition of dilute HCl. After 28 days, the NOEC and LOEC for survival were 41.3 and 90.9 mg/kg d. w., respectively. With regard to the amphipod length and total biomass, the 28 days NOEC and 28 days LOEC were 90.9 and 200 mg/kg d. w., respectively.

### 4.2.4 Additional Tests on Feeding Mode and Ammonia Formation

Over the course of the aforementioned testing of MDA with benthic organisms, the potential confounding of results from feeding of the organisms, the feed types used, and potential for associated ammonia formation from decay of the food was questioned. Additional studies were conducted to improve the understanding of how these test parameters may have influenced results and interpretations of the aforementioned tests. West et al. (2004) reported that spiking of sediment with nettle powder (*Urtica* sp.) at the beginning of formulated sediment tests results in a stronger and earlier increase in ammonia concentration than feeding via the overlying water with semi-continuous addition of TetraMin®. The discovery of these feeding-related influences on ammonia production prompted further investigation using a factorial design study to investigate the influence of different feeding parameters on the water quality in sediment tests with pre-spiked sediments (Egeler and Gilberg 2005c). The food spiked into the sediment was either Urtica: cellulose = 1:1 (wt:wt) or cereal leaves (wheat; *Triticum aestivum*), test-substance (4,4′-MDA, 97.9 %) was added or not (10 mg/kg d. w.), aqueous medium was either M4 (Elendt medium) or a pH-reduced medium (by reduction of $NaHCO_3$), and organisms added were either the blackworm *L. variegatus* or the amphipod *H. Azteca*. The tests were run for 28 days. For *L. variegatus*, cereal leaves was shown to be the best food source concerning total number of worms and biomass. The urtica powder/cellulose combination food source resulted in a significant reduction of biomass, especially when the M4 medium was modified. For *H. azteca*, survival and length were not influenced by any of the feeding factors evaluated, but the biomass was significantly reduced when the modified M4 medium was used. The ammonia concentration in the pore water and overlying water was significantly higher when cereal leaves were used as food source. The presence of MDA had no measureable influence on the ammonia content. This can be regarded as an indication that nitrification is not inhibited by up to 10 mg/kg d. w. MDA in the formulated sediment.

In the previously-described second test with the blackworm *L. variegatus* (Egeler 2002), pH values were taken at day 1 and 28 and found both to be below pH 8. However, pH values were not recorded in between and ammonia was not monitored. In sediment spiked with nettle powder, the pH exceeded a value of 8 after 48 h, and then declined to levels below 8 after 11 day, influenced by addition of diluted HCl. Ammonia, however, achieved the highest concentration at day 4 of the experiment (West et al. 2004). The authors further demonstrated that ammonia formation is less a problem in case of semi-continuous feeding with TetraMin®. As a result, the lower NOEC and LOEC in the second study with *L. variegatus* are probably attributable to interference by ammonia. Although the number of worms with new posterior and anterior end seems to be most sensitive parameter in the second study, results from the first study demonstrated that this endpoint suffers from a high variability even in the control. As a result, this endpoint is not suitable for the interpretation of the toxicity of MDA to *L. variegatus* under the test design employed.

Sediment tests performed with MDA generally showed a high variability in outcomes. This was at least partly attributable to the confounder ammonia production. In addition, MDA can be regarded as a "difficult substance" for sediment testing. For example, in an OECD Guideline 308 study simulating the fate of radiolabeled MDA in a surface water and sediment system (Schaefer and Ponizovsky 2013), there was no MDA detectable in the aquatic layer 7 days after test initiation. After 100 day, 5.7 % of the MDA was transformed to $CO_2$, about 6 % MDA metabolites were detectable in the water, 12 % MDA products were extractable from the sediment and 90 % remained unextractable (retrievability about 114 %). Especially in the first few days after addition to the sediment, the MDA suffers rapid transformations against the time frame of the chronic sediment tests. The parent substance and the metabolites are assumed to show different toxicity against the sediment organisms. The design as used was chosen as such because it is not known whether MDA itself or one of the transformation products is most critical. The variability in test outcomes observed seems to be attributed to subtle variations in conditions of these dynamic test systems.

## 4.3 *Toxicity to Terrestrial Organisms*

The reactivity of the MDAS with natural organic matter indicates the potential for their association with surface soils should direct exposures occur as a result of spillages, or indirect exposures such as a hypothesized hydrolysis and deposition of atmospheric emissions of MDI substances (Government of Canada 2014). Data with soil organisms are summarized in Table 5.

### 4.3.1 Tests with Soil Bacteria

Soil bacterial serve a critical role in the cycling of nitrogen in the environment. The decay of nitrogen-containing organic matter can result in the release of ammonia, and this ammonia can then be oxidized by specific nitrifying bacteria to nitrite and nitrate. Because several aromatic amine substances were shown to behave as inhibitors of these nitrifying bacteria in soil (Zhang et al. 2010), the potential for MDAS to have similar effect was evaluated using a modification to the OECD Guideline 216 (2000). The standard version of this OECD test recommends that plant material is added to soil and to check for the formation of nitrate. The principle of this test is that as heterotrophic soil bacteria decompose this plant material to release ammonia, the amount of nitrate formed from this ammonia is dependent upon the activity of the specific denitrifying bacterial population in the soil. A reduction in the rate or total mass of nitrate production in soil treated with the test substance, compared to that in an untreated control soil, is interpreted as

**Table 5** Summary of acute and chronic toxicity studies for 4,4′-methylenedianiline (MDA) with soil organisms

| Organism | Endpoint | Value (mg/kg d. w.) | References |
|---|---|---|---|
| Oats *Avena sativa* | 17-days growth | $EC_{50} = 353$ | Van der Hoeven et al. (1992a) |
| Lattuce *Lactuca sativa* | 17-days growth | $EC_{50} = 128$ | |
| Radish *Raphanus sativus* | 6-days emergence | $EC_{20} = 96^{a}$ | Kim et al. (2002) |
| | | $EC_{50} = 331^{a}$ | |
| Earthworm *Eisenia fetida* | 14 days mortality | $EC_{50} = 444$ | Van der Hoeven et al. (1992b) |
| Earthworm *Eisenia fetida* | 56 days reproduction/ mortality | $EC_{50} = 333$ | Moser and Hamberge (2012) |
| | | $EC_{10} = 11$ | |
| | | LOEC = 18 | |
| Springtail *Folsomia candida* | Reproduction 28 days | NOEC = 562 | Moser and Schott (2011) |
| | | LOEC = 1000 | |
| Bacteria Nitrification in soil | $NH_4^+$ depletion; $NO_3^-$ formation | NOEC = 1000 | Schwarz (2013) |
| Nitrification in activated sludge | $NH_4^+$ depletion | $IC_{50} = 61$ mg/L | |

[a]In water

potential inhibition of these nitrifying soil bacteria. However, recognizing that the MDAS are reactive aromatic compounds, and nitrite, formed transiently during nitrification of ammonia, may react directly with the MDAS by forming azo-compounds or nitroso-aromatics. If such reactivity were to occur under conditions of the standard OECD 216 test, the resulting reduction in nitrate formation may be mis-interpreted as inhibition of nitrifying bacteria. In a screening study to examine potential for this reactivity, (unpublished data of BASF Polyurethanes, Gericke 2010) the stability of 4,4′-MDA was observed in an aquatic solutions at pH = 6 in the presence and absence of equimolar amounts of sodium nitrite; the decay of nitrite was not investigated. The solutions were incubated at about 22 °C in the dark. The results of this screening test, shown in Fig. 10, indicated an instability (reactivity) of 4,4′-MDA in the presence of nitrite under these slightly acidic conditions. In the presence of nitrite, the initially colorless 4,4′-MDA solution became yellow after 7 days.

With this knowledge, an OECD Guideline 216 study was performed in a modified way (Schwarz 2013), where instead of plant material the soil was spiked with about 75 mg ammonium sulfate per 5 g dry soil. Over the test period, both the decline of ammonium and the formation of nitrate were monitored. The soil was spiked with 0, 10, 20, 50, 125, 250, 500 or 1000 mg 4,4′-MDA per kg soil (d. w.). The $EC_{10}$ value was shown to be above 1000 mg/kg after 14 and 28 days in terms of ammonia depletion, as well as in terms of nitrate formation. Only after 7 days, the $EC_{10}$ was between 500 and 1000 mg/kg, indicating a potential minor and transient

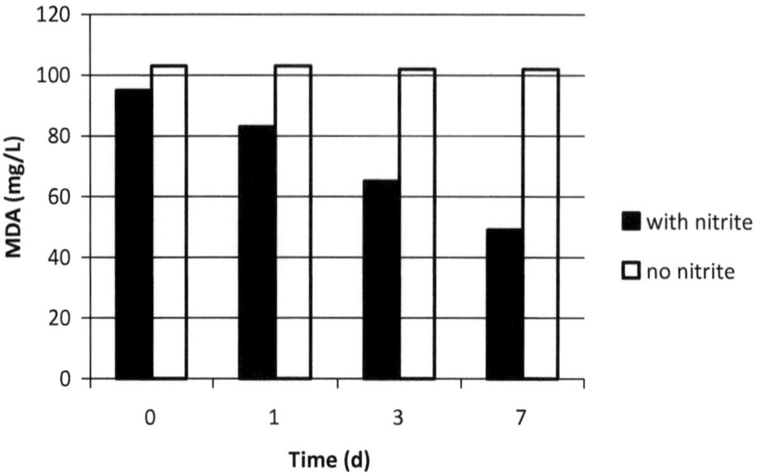


**Fig. 10** Time-dependent concentrations of 4,4′-MDA in aqueous solution in the presence (*black bars*) and absence (*white bars*) of nitrite at pH = 6

effect on the soil bacteria. The amount of MDA measured in aqueous soil extracts taken for nitrate analyses was shown to rapidly decline and was about 14–60 % of the initial lowest and highest MDA concentrations tested, respectively, at the beginning of this test. At the end of the test (28 days), recoveries of the MDA were < 10 % of their initial nominal values. This result is in line with observations in the previously discussed sediment tests with MDA, where an irreversible absorption of MDA to soil organic matter along with biodegradation are the most likely explanation disappearance from the test systems. So, a limited inhibition of nitrification was observable only at the beginning at the highest soil load (1000 mg/kg d. w.), when some freely available MDA was likely still available. The quantitative formation of nitrate further indicates that transiently-formed nitrite was not significantly reacted with MDA, as was shown to be the case in pH 6 aqueous solution (Gericke 2010).

The ISO 9509 test (2006) can be used to check the nitrification inhibition in activated sludge. Initial cursory tests were inconclusive as 17 mg/L seemed to be the $IC_{50}$ concerning ammonium decay, but it was difficult to discern even a 10 % inhibition of nitrate formation (Schupp 2014). Therefore, a definitive test was performed with activated sludge of the municipal STP of the city Steinfurt, Germany. The sludge (3 g solids per liter, dry wt.) was spiked with ammonium sulfate and 0, 9.4, 18.7, 47 and 93.5 mg/L 4,4′-MDA (98 % pure). The N-Allyl-thiourea substance, a known inhibitor of nitrification, served as control. The nutrient solution contained 2.71 g/L $NH_2SO_4$ and 5.47 g/L $NaHCO_3$ and was diluted 1:10 for the test runs. Tests were run in duplicate. After 4 h incubation at 20 °C, samples were taken, filtered through a 0.46 μm membrane, and ammonium ions were analyzed photometrically after reaction with $K_2HgI_4$. A blank experiment proved that MDA does not interfere with this photometrical method. In this test, MDA shows moderate

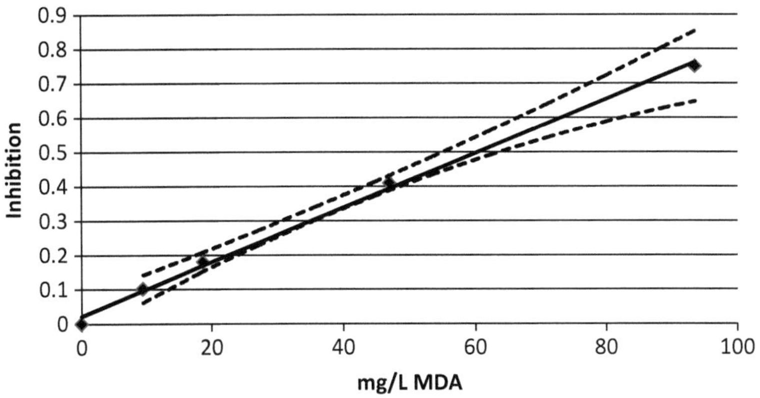


**Fig. 11** Inhibition of ammonium nitrification by MDA in an ISO 9509 test

nitrification inhibition with $IC_{50} = 61$ mg/L (Fig. 11). HPLC-UV analysis of the 47 mg/L solution after the test showed that MDA was neither absorbed on sludge nor did it suffer other routes of decay.

### 4.3.2 Toxicity to Terrestrial Plants

The potential effect of MDA on the emergence and growth of monocot (oat, *Avena sativa*) and dicot (lettuce, *Lactuca sativa*) vascular plants was investigated according to OECD Guideline 208 (Van der Hoeven et al. 1992a). The purity of the test substance was 99.5 %. Oat was sown at 1 cm depth, and lettuce on the surface of a semi-natural soil. The soil was previously spiked with sand that was coated with MDA. For the latter, MDA was dissolved in acetone, the sand soaked with this MDA-solution, and the acetone was then evaporated under a stream of nitrogen. Tests were run for 17 days and five replicates at ten seeds per concentration (0, 3.2, 10, 32, 100, 320 and 1000 mg/kg d. w. soil, nominal concentrations). Short-term endpoints were associated with concentrations at which no effects were observed, and the lowest concentration at which effects were observed by pair-wise binomial test ($2 \times 2$ contingency table).

For the emergence endpoint, concentrations were defined as 320 and 100 mg/kg d. w. at which no effects were observed for 17 days exposures to oat and lettuce, respectively. For growth, the corresponding concentrations showing no effects were 100 and 10 mg/kg d. w. for oat and lettuce, and the corresponding $EC_{50}$ values were 353 and 128 mg/kg d. w., where these endpoint values are based on nominal concentrations. The dicotyledonous lettuce, therefore, is somewhat more sensitive to 4,4′-MDA than the monocotyledonous oat.

In another seed germination test, Kim et al. (2002) investigated the influence of 4,4′-MDA dissolved in water and poured onto filter paper in petri-dishes on the germination of radish seeds. The precise species, not mentioned in the paper, probably was *Raphanus sativus*. After 6 days at 20 °C, germination was assessed visually. The reliability of this test may be questioned, as the authors claim to have used MDA solutions up to 1 % MDA (10 g/L) which is clearly beyond the water solubility of MDA (Table 2). Data were presented in graphs only, and germination is estimated to be 86, 48, 29, 14, 5 and 0 % at 0.01, 0.02, 0.06, 0.1, 0.5 and 1.0 % MDA in water, respectively. If the two highest concentrations are omitted due to potential insolubility of MDA, the 6 days $EC_{50}$ is interpreted to be 331 mg/L, and a 6 days $EC_{20}$ is 96 mg/L is estimated by logit analysis.

### 4.3.3 Toxicity to the Earthworm *Eisenia fetida*

Van der Hoeven et al. (1992b) investigated the acute toxicity of 4,4′-MDA against *E. fetida* after 14 days exposure according to OECD Guideline 207 (1984), using nominal test concentrations of 0, 18, 32, 56, 100, 320 and 560 mg/kg d. w. applied to five replicates per concentration. The 4,4′-MDA (99.5 % purity) was dissolved in acetone and added to artificial soil. Subsequently, the acetone was evaporated under a stream of nitrogen. The 14 days $LC_{50}$ was determined to be 444 mg/kg d. w.

Moser and Hamberge (2012) investigated the effect of 4,4′-MDA on the reproduction of *E. fetida* according to OECD Guideline 222 (2004) and in compliance with GLP. Stock solutions of MDA (purity 98 %) were prepared in ethanol, and the concentrations were verified by HPLC-UV. Defined volumes of the MDA solutions were added to quartz sand, and the ethanol was evaporated. The coated quartz sand was used incorporated into the artificial soil, containing 5 % sphagnum peat. Nominal concentrations were 18 32, 56, 100, 180 and 320 mg/kg d. w., with four replicates used at each concentration and ten worms per replicate. Worms were fed with cow manure which was free of contamination and veterinary pharmaceuticals. At day 56, test vessels were analyzed. For the mortality endpoint, the 56 days NOEC was 180 mg/kg d. w., the 56 days $EC_{10}$ was 92.06 mg/kg d. w., and the 56 days $EC_{50}$ was 333 mg/kg d. w. soil. For adult biomass endpoint, the 56 days NOEC and 56 days LOEC values were 32 and 56 mg/kg d. w., respectively. For reproduction, the 56 days LOEC was 18 mg/kg d. w. As no NOEC (reproduction) could be derived the 56 days $EC_{10}$ of 11.2 mg/kg (95 % C.I.: 0.6–21.3 mg/kg d. w.) was statistically determined instead.

### 4.3.4 Toxicity to the Collembolan Species *Folsomia candida*

The effect after 4 weeks of exposure to 4,4′-MDA on the collembolan species *F. candida* was investigated by Moser and Schott (2011) according to OECD Guideline 232. The 4,4′-MDA test substance (98 % pure), dissolved in ethanol, was spiked to artificial soil containing 5 % sphagnum peat. Spiking solutions were analyzed by HPLC-UV. Nominal concentrations were 56.2, 100, 178, 316, 562 and 1000 mg/kg d. w. soil. For adult mortality, there was no clear dose–response; the highest mortality of 20 % was observed at 316 mg/kg; for reproduction, the 28 days

NOEC was 562 mg/kg and the 28 days LOEC was 1000 mg/kg d. w. The reference substance boric acid generated the expected results.

### 4.3.5 Acute Toxicity to Birds

Hurlbut et al. (1983) report an acute $LD_{50}$ of 148 mg/kg for redwinged blackbirds (*Agelaius phoeniceus*), fed with grains containing 4,4′-MDA. Data on purity of the substance and stability of MDA in/on the food are not provided, and many experimental details are lacking. The reliability of the data generated, therefore, is questionable.

## *4.4 Potential for Endocrine-Modulating Effects and Brief Overview of the Toxicity to Mammalia*

A review of data on endocrine activity of MDA was issued by Jaeger and Collins (2011) and Jaeger et al. (2012). A summary of the reviewed findings is presented as follows:

In yeast receptor binding studies, 4,4′-MDA and pMDA showed neither estrogenic nor androgenic activity and no anti-estrogenic activity in the dose range of $10^{-7}$ to 0.1 mmol/L; however, 4,4′-MDA and pMDA showed apparent anti-androgenic activity at concentrations of $10^{-3}$ and $10^{-2}$ mmol/L, respectively (Kolle and Landsiedel 2010a, b, c, d). Since MDA has potential to react directly with the endogenous hormone (testosterone) and with proteins which are the construct of the androgen receptor, the apparent anti-androgenic activity associated with high MDA concentrations could be an artifact of this reactivity and not necessarily indicative of receptor-mediated antagonism.

In a subacute oral study, castrated female rats received 150 and 200 mg/kg MDA for up to 14 days (Tullner 1960). The body weight decreased whereas uterine, thyroid and adrenal weights increased. The adrenals showed extensive lipid accumulation, and thyroid follicles contained little or no colloid.

Intravenous administration of 50 and 100 mg/kg MDA caused a transient reduction in 17-hydroxycorticoid output in the dog (Tullner 1960).

In ovarectomized rabbits, subcutaneous administration of MDA caused a progestational response. However, in ovarectomized and adrenalectomized rabbits, MDA did not cause a response, whereas subcutaneous progesterone did. Cortisol did suppress the progestational response to MDA, but not that due to progesterone (Tullner 1960).

Rats received 0, 80, 400 or 800 ppm MDA in drinking water for 3 months (Ciba-Geigy 1982). Besides hematotoxic and hepatotoxic effects, thyroid follicular epithelial cells were hypertrophic, and the glandular structures showed a diffuse hypertrophy and colloid depletion. In another subchronic drinking water study with rats, similar effects were reported (DHHS NTP 1983). In both studies, gonads and accessory organs were not affected.

In mice, MDA caused mainly hepatotoxicity; goiters with papillary hyperplasia and vacuolization of the colloid were observed in one high dose male and female animal (DHHS NTP 1983).

In chronic drinking water studies, MDA caused thyroid follicular cell adenomas and carcinomas in rats and mice (DHHS NTP 1983). This effect might be attributable to irreversible inhibition of thyroid peroxidase by the reactive MDA substance at high dose levels. In a later study, MDA was shown to be an effective inhibitor of hog thyroid peroxidase (TPO) in vitro with an IC50 of 0.06 μM (Freiberger 1994). However, direct genotoxicity of MDA in thyreocytes was demonstrated by a positive comet assay (Martelli et al. 2002).

## 5 Summary

Concerning chronic toxicity, *D. magna* is the most sensitive species tested against MDA aquatic exposures, with a 21 day-NOEC of 0.00525 mg/L. Exposure of daphnids takes place via the aquatic phase. Other species of the same phylum (*Arthropoda*) appear to be less sensitive albeit with exposures via soil or sediment, with a 28 days-NOEC of 562 mg/kg d. w. soil (*F. candida*) and 41.3 mg/kg d. w. sediment (*Hyalella azteca*), for reproductive and survival endpoints, respectively. Also for acute toxicity, *D. magna* is more sensitive than the other species, with an 48 h-EC50 that spreads over two orders of magnitude, ranging from 0.019 to 2.7 mg/L; the reason for this large difference is not clear at the moment. Data laying at both ends of the range are from the same lab, testing the same charge of MDA, and the only difference visible so far is the feeding mode. That is, daily feeding with algae resulted in an 48 h $EC_{50}$ of 0.019–0.06 mg/L, whereas the 48 h $EC_{50}$ from the acute test with no feeding is 2.47 mg/L. Future tests might investigate whether intake of MDA via the food chain results in a significantly higher body burden in the daphnids than intake via water, only. Fish show a more uniform reaction to MDA, with 96 h-LC50 values ranging from about 20 to 60 mg/L; chronic data for fish are not available. Acute toxicity data for algae and cyanobacteria are in the range of 1–10 mg/L; based on growth-rate, the 72 h-NOEC or $EC_{10}$ of MDA to algae is 0.3–9.3 mg/L.

For sediment organisms, the blackworm *L. variegatus* shows the highest sensitivity to MDA with NOEC values between $\leq 3.75$ mg/kg and 30 mg/kg d. w., followed by the amphipod *H. azteca*. The higher sensitivity of *L. variegatus* in the second study compared to the first study is obviously attributable to the different feeding regimes (semi-continuous feeding against pre-spiked sediment). One argument might be that semi-continuous feeding allows the organisms to avoid the contaminated food. However, a change from semi-continuous feeding to sediment pre-spiked with nettle powder (*Urtica* sp.) results in an earlier and much stronger increase in ammonia concentration in the system. This became apparent after both studies on *the blackworm* were finalized. The ammonia 96 h-EC50 for the *blackworm* is 0.69 mg/L at pH = 8.2, and the 96 h-EC10 at pH = 8.2 is 0.33 mg/L (Hickey and Vickers 1994). As a result, the lower NOEC and LOEC in the second study with *L. variegatus* are probably attributable to interference by ammonia.

MDA binds irreversibly to soil and sediment which may explain the general, but not uniform lower sensitivity of soil and sediment organisms against aquatic organisms. The intrinsic toxicity of MDAS, beyond that attributed to a baseline narcosis mode of action, is imparted by the reactivity of the primary aromatic amine groups. Thus, reaction of these amine groups with organic matter (macromolecules) would be expected to coincide with reduced reactivity, toxicity, and bio-availability However, species with intense soil or sediment contact (*L. variegatus* and *E. fetida*) show in general lower NOEC values than those organisms with less direct contact (3.75 and 11 mg/kg d. w., respectively). On the one hand it may be hypothesized that this intense contact to soil-bound MDA is one reason for the higher sensitivity; on the other hand, metabolic capacity against MDA of the organisms tested is unknown at this point in time and might as well explain differences in species sensitivity. For plants there are only acute data available; for the terrestrial species tested, *L. sativa* is more sensitive to MDA than *E. fetida*.

Limited aquatic data available so far do not indicate that the toxicity of pMDA is different from that of MDA. In addition, the limited set of data generated with the marine *M. macrocopa* (crustacean, acute and chronic test), *N. fustulum* (diatom) and *V. fisheri* (bacteria) do not indicate that sea water organisms are more sensitive to MDA than fresh water organisms.

In mammals, MDA is unlikely to interact directly with the endocrine-mediated homeostatic, growth, and developmental pathways; interaction with the adrenergic system cannot be ruled out, and apparent effects of MDA on the thyroid hormone system have been demonstrated. MDA inhibits the thyroid peroxidase enzyme which might contribute to the thyroid gland tumors observed in chronic studies with rats and mice. Some anti-androgenic activity in vitro did not prevail in the in vivo studies with rats and mice. Main target organs in mammals are the liver and the thyroid gland. MDA shows genotoxicity in different test systems and is a carcinogen by GHS category 1b.

## 6 Conclusion

The toxicity of 4,4′-Methylenedianiline, its isomers and homologues (MDA substances; MDAS) against organisms in the environment was iteratively tested over at least four decades. This review underlines the importance of accurate substance identification and analyses that accompany ecotoxicity tests for substances which can react with or in environmental compartments. Especially for the sediment and soil compartment that absorb MDA reversibly and irreversibly, the influence of these processes on the biological availability and toxicity of MDA and its products requires careful consideration. Sediment tests with MDA were performed at a time where the increase in knowledge concerning reactive substance was on a steep slope, and the guideline for *L. variegatus* was under development. This, together with the dynamics of the MDA molecule in the sediment, may explain variations in findings of the sediment tests. There is a remarkably high variation in toxicity values in acute daphnia studies. Whereas most values are in the range of 0.46–5.7 mg/L which represent the typical variation in ecotoxicology there is one study showing significant lower EC50 values in the range of 0.019–0.06 mg/L, thus indicating that the factors governing ecotoxicity of MDAS to daphnids is currently not fully understood.

**Acknowledgement** This work was sponsored by the International Isocyanate Institute, Inc. The views presented in this paper are those of the authors and not necessarily those of the sponsor.

In addition we would like to acknowledge the useful input and translation of the summary provided by S. Shen (Dow Chemical (China) Investment Limited Company, Shanghai, China).

*Conflict of interest*: T. Schupp worked for BASF, a MDA-producer, until 2012; H. Allmendinger is a consultant for Currenta GmbH & Co. OHG; B.T.A. Bossuyt is working for Huntsman, a MDA-producer; B. Hidding is working for BASF, a MDA-producer; B. Tury is a consultant for International Isocyanates, Inc.; R.J. West is working for Dow Chemical Company, a MDA-producer.

## References

Alport DC, Gilbert DS, Outterside SM (eds) (2003) MDI and TDI: safety, health, and the environment—a source book and practical guide. Wiley, West Sussex, England

Bringmann G, Meinck F (1964) Wassertoxikologische Beurteilung von Industrieabwässern. Gesund Ing 85(8):229–236

Carvajal-Diaz J (2015) IHS chemical economics handbook: aniline. IHS Inc, Englewood, CO, USA

Caspers N, Hamburger B, Kanne R, Klebert W (1986) Ecotoxicity of toluenediisocyanate (TDI), diphenylmethanediisocyanate (MDI), toluenediamine (TDA), diphenylmethanediamine (MDA). International Isocyanate Institute, Manchester, UK, III Report No. 10417

Caspers N, Mueller N (1992) Untersuchungen zum oekologischen Verhalten von Phenylbase MDA 70. Bayer Pharma AG, Wuppertal, Germany, Pruefnummer: 281 A/91

Ciba-Geigy (1982) MDA: three months toxicity study in rats (drinking water). Ciba-Geigy Ltd, Basle, Switzerland, Finalized report TK 10504

CITI (1992) 4-Methylphenylene-1,3-diamine (CAS-No. 95-80-7) & 4,4-′-diaminodiphenylmethane (CAS-No. 101-77-8). [Acute toxicity, biodegradability and bioaccumulation data]. In: Biodegradation and bioaccumulation data of existing chemicals based on the CSCL Japan. Chemical Inspection & Testing Institute Japan, Japan Chemical Industry Ecology-Toxicology & Information Center, Tokyo, pp 3–24 & 4–6. ISBN: 4-89074-101-1

Colón D, Weber EJ, Baughman GL (2002) Sediment-associated reactions of aromatic amines. 2. QSAR development. Environ Sci Technol 36:2443–2450

Cowen WF, Gastinger AM, Spanier CE, Buckel JR, Bailey RE (1996) Sorption and microbial degradation of tuloene diamines and methylene dianiline in soil under aerobic and anaerobic conditions. International Isocyanate Institute, Manchester, UK, III Report No. 11230

Cowen WF, Gastinger AM, Spanier CE, Buckel JR, Bailey RE (1998) Sorption and microbial degradation of toluenediamines and methylendianiline in soil under aerobic and anaerobic conditions. Environ Sci Technol 32:598–603
DHHS NTP (1983) NTP technical report on the carcinogenesis studies of 4,4′-methylenedianiline dihydrochloride (CAS No. 13552-44-8) in F344/N rats and B6C3F1 mice (drinking water studies). US Department of Health and Human Services, National Toxicology Program. PB83-238824; NTP-TR-248; NIH Publication no. 83-2504. Research Triangle Park, NC
ECHA (2014a) Formaldehyde, oligomeric reaction products with aniline [REACH Dossier]. European Chemicals Agency, Helsinki, Finland, http://echa.europa.eu/registration-dossier/-/registered-dossier/14114. Accessed 21 June 2014
ECHA (2014b) 4,4′-Methylenedianiline [REACh Dossier]. European Chemicals Agency, Helsinki, Finland, http://echa.europa.eu/registration-dossier/-/registered-dossier/15201. Accessed 21 June 2014
Egeler P, Ginzburg N (2001) The toxicity of 4,4′-methylenedianiline (MDA) to *Lumbriculus variegatus*. International Isocyanate Institute, Manchester, UK, III Report No. 11436. Robust study summary available at: http://echa.europa.eu/registration-dossier/-/registered-dossier/15201/6/3/?documentUUID=cac3546d-bce8-469f-8fb8-9ec304e91916. Accessed 13 May 2015
Egeler P (2002) The toxicity of MDA to *Lumbriculus variegatus*: second study. International Isocyanate Institute, Manchester, UK, III Report No. 11459. Robust study summary available at: http://echa.europa.eu/registration-dossier/-/registered-dossier/15201/6/3/?documentUUID=fc0116d2-f5e1-4510-970a-766bcde6ec6f. Accessed 13 May 2015
Egeler P (2004) Ammonium chloride: a study on the toxicity to the sediment dweller *Hyalella azteca*. International Isocyanate Institute, Manchester, UK, III Report No. 11499
Egeler P, Gilberg D (2005a) The toxicity of MDA to *Chironomus riparius*. International Isocyanate Institute, Manchester, UK, Report No. 11517. Robust study summary available at: http://echa.europa.eu/registration-dossier/-/registered-dossier/15201/6/3/?documentUUID=3b01e515-e9ce-457b-901e-cf317c876e59. Accessed 13 May 2015
Egeler P, Gilberg D (2005b) The toxicity of MDA to *Hyalella azteca*. International Isocyanate Institute, Manchester, UK, III Report No. 11518. Robust study summary available at: http://echa.europa.eu/registration-dossier/-/registered-dossier/15201/6/3/?documentUUID=a5ed19f2-a4f9-48cf-8a9c-53a06392ef81. Accessed 13 May 2015
Egeler P, Gilberg D (2005c) The optimisation of feeding conditions in sediment tests. International Isocyanate Institute, Manchester, UK, III Report No. 11510
DIN (1980) German Standard Methods for the Examination of Water, Waste Water And Sludge—bio-assays (Group L, No. 38412–12), determination of the effect of waste water and industrial effluences on fish, fish-test (l20). German Institute for Standardization, Berlin, Germany
European Union (2001) 4,4′-Methylenedianiline, CAS-No. 101-77-9, EINECS-No. 202-974-4. European Union risk assessment report, volume 9. Office for Official Publications of the European Union, Luxembourg. ISBN 92-894-0484-1
Freiberger A (1994) Irreversible inhibition of thyroid peroxidase (TPO) by thyreotoxic aromatic amines in vitro. Naunyn-Schmiedeberghs Arch Pharmacol 349(S6):Abs440
Fujiwara K (1981) Aquatic life study phase II, part 2. Accumulation of MDI, TDI, MDA and TDA in fish and their toxicity. International Isocyanate Institute, Manchester, UK, III Report No. 10019
Fujiwara K (1982) Studies on the effect of TDA and MDA on plankton—Daphnia. International Isocyanate Institute, Manchester, UK, III Report No. 10118, Robust study summary available at: http://echa.europa.eu/registration-dossier/-/registered-dossier/15201/6/2/4. Accessed 13 May 2015
Gericke G (2010) Stabilitaet von MDA in Gegenwart von Nitrit bei pH=6. Analysenbericht A20102055. Elastogran GmbH, Elastogranstrasse 60, D-49448 Lemfoerde, Germany
Government of Canada (2014) Draft screening assessment for methylenediphenyl diisocyanates and methylenediphenyl diamines. Government of Canada, Chemicals Management Plan

Division, Gatineau, Quebec, Canada. http://www.chemicalsubstanceschimiques.gc.ca/group/diisocyanate-eng.php

Hickey CW, Vickers ML (1994) Toxicity of ammonia to nine native New Zealand freshwater invertebrate species. Arch Environ Contam Toxicol 26:292–298

Hurlbut J, Schafe EW, Bowle WA (1983) The acute oral toxicity, repellency, and hazard potential of 998 chemicals to one or more species of wild and domestic birds. Arch Environ Contam Toxicol 12:355–382

ISO 9509 (2006) Water quality—toxicity test for assessing the inhibition of nitrification of activated sludge microorganisms (ISO 9509:2006). http://www.beuth.de/de/norm/din-en-iso-9509/89012595. Accessed 11 Mar 2016

Jaeger R, Collins MA (2011) Assessment of MDA effects from toxicity studies with regard to endocrine modulation. Global Isocyanates Ltd, Manchester, UK, GIL Report No. 2011/C

Jaeger R, West RJ, Collins MA (2012) Assessment of MDA effects from toxicity studies with regard to endocrine modulation. Naunyn-Schmiedeberghs Arch Pharmacol 385(S1):Abs. 176

Kaiser KLE, Palabrica VS (1991) *Photobacterium phosphoreum* toxicity data index. Water Pollut Res J Can 26(3):361–431

Kim M-N, Jang J-C, Lee I-M, Lee H-S, Yoon J-S (2002) Toxicity and biodegradation of diamines. J Environ Sci Health B37(1):53–64

Kolle SN, Landsiedel R (2010a) Oligomeric MDA: screening for potential androgenic and anti-androgenic activity. International Isocyanate Institute, Manchester, UK, III Report No. 11597

Kolle SN, Landsiedel R (2010b) Oligomeric MDA: screening for potential estrogenic and anti-estrofiggenic activity. International Isocyanate Institute, Manchester, UK, III Report No. 11598

Kolle SN, Landsiedel R (2010c) 4,4′-MDA: screening for potential androgenic and anti-androgenic activity. International Isocyanate Institute, Manchester, UK, III Report No. 11603

Kolle SN, Landsiedel R (2010d) 4,4′-MDA: screening for potential estrogenic and anti-estrogenic activity. International Isocyanate Institute, Manchester, UK, III Report No. 11604

Li H, Lee LS, Jafvert CT, Graveel JG (2000) Effect of substitution on irreversible binding and transformation of aromatic amines with soils in aqueous systems. Environ Sci Technol 34:3674–3680

MacNab JI (1999) Determination of physical properties of 4,4′-MDA. International Isocyanate Institute, Manchester, UK, III Report No. 11330. Robust study summary available at: http://echa.europa.eu/registration-dossier/-/registered-dossier/15201/6/2/4. Accessed 13 May 2015

Martelli A, Carrozzino R, Mattioli F, Brambilla G (2002) DNA damage induced by 4,4-′-methylenedianiline in primary cultures of hepatocytes and thyreocytes from rats and humans. Toxicol Appl Pharmacol 182(3):219–225

Mitsubishi Chemical Safety Institute (2008a) Acute toxicity of 4,4′-MDA to *Selenastrum capricornutum*. International Isocyanate Institute, Manchester, UK, III Report No. 11545. Robust study summary available at: http://apps.echa.europa.eu/registered/data/dossiers/DISS-9c7b34e3-c5c4-7414-e044-00144f67d249/AGGR-acfea30d-ad97-4274-b8ce-aabd18d556da_DISS-9c7b34e3-c5c4-7414-e044-00144f67d249.html. Accessed 13 May 2015

Mitsubishi Chemical Safety Institute (2008b) Acute toxicity of 4,4′-MDA to *Daphnia magna*. International Isocyanate Institute, Manchester, UK, III Report No. 11546. Robust study summary available at: http://echa.europa.eu/registration-dossier/-/registered-dossier/15201/6/2/6. Accessed 13 May 2015

Mitsubishi Chemical Safety Institute (2008c) Chronic toxicity of 4,4′-MDA to *Daphnia magna*. International Isocyanate Institute, Manchester, UK, III Report No. 11547. Robust study summary available at: http://echa.europa.eu/registration-dossier/-/registered-dossier/15201/6/2/5. Accessed 13 May 2015

Mitsubishi Chemical Safety Institute (2008d) Acute toxicity of 4,4′-MDA to *Oryzias latipes*. International Isocyanate Institute, Manchester, UK, III Report No. 11548. Robust study summary available at: http://echa.europa.eu/registration-dossier/-/registered-dossier/15201/6/2/2/?documentUUID=194ccb6b-2b28-4265-a6f8-d65c93a0b53b. Accessed 13 May 2015

Moser T, Hamberge R (2012) 4,4′-MDA: reproduction toxicity to the earthworm *Eisenia fetida*. International Isocyanate Institute, Manchester, UK, III Report no. 11618. Robust study summary available at: http://echa.europa.eu/registration-dossier/-/registered-dossier/15201/6/4/2/?documentUUID=3b463e27-4e1b-46ad-9301-5c0b097066a9. Accessed 13 May 2015

Moser T, Schott C (2011) 4,4′-MDA: toxicity to the collembolan species *Folsomia candida*. International Isocyanate Institute, Manchester, UK, III Report No. 11614. Robust study summary available at: http://echa.europa.eu/registration-dossier/-/registered-dossier/15201/6/4/3. Accessed 13 May 2015

Munk R, Kirsch P (1988) Golden orfe (*Leuciscus idus L.*, golden variety): report on the study of the acute toxicity. Project No. 10F0621/875279. BASF AG, Ludwigshafen, Germany. Robust study summary available at: http://echa.europa.eu/registration-dossier/-/registered-dossier/15201/6/2/2/?documentUUID=af1adb0e-7097-4248-a158-c5d263c6c040. Accessed 13 May 2015

OECD (1981) OECD guidelines for testing of chemicals, guideline no. 112: dissociation constants in water. Organization for Economic Cooperation and Development, Paris, France, Adopted 12 May 1981

OECD (2004) OECD guidelines for testing of chemicals, draft guideline no. 218: sediment-water chironomid test using spiked sediment. Organization for Economic Cooperation and Development, Paris, France, Adopted 13 Apr 2004

Parris GE (1980) Covalent binding of aromatic amines to humates. 1. Reactions with carbonyls and quinones. Environ Sci Technol 14(9):1099–1106

Rhône-Poulenc Chimie (1977) Biological action of MDI and TDI in water. International Isocyanate Institute, Manchester, UK, III Report No. 10092

Riedhammer C, Schwarz-Schulz B (2001) The newly proposed EU risk assessment concept for the sediment compartment. J Soils Sediments 1(2):105–110

Rufli H, Mueller K (1985a) Report on the test for acute toxicity of TK 10504 to algae. (OECD guideline no. 201, Paris 1984). Ciba-Geigy Ltd, Basle, Switzerland, Project No. 850731. Robust study summary available at: http://echa.europa.eu/registration-dossier/-/registered-dossier/15201/6/2/6/?documentUUID=a115b7b8-fa62-405b-86e1-cfa39aefab56. Accessed 13 May 2015

Rufli H, Mueller K (1985b) Report on the test for acute toxicity of TK 10504 to *Daphnia magna*. OECD guideline no. 202 Paris (1984). Ciba-Geigy Ltd, Basle, Switzerland, Robust summary available at: http://echa.europa.eu/registration-dossier/-/registered-dossier/15201/6/2/4/?documentUUID=ca31cdd6-96f7-4ac9-90ed-d22bf43d92ad. Accessed 11 March 2016

Rufli H, Mueller K (1985c) Report on the test for acute toxicity of TK 10504 to zebra fish. OECD guideline no. 203 Paris (1984). Ciba-Geigy Ltd, Basle, Switzerland, Robust study summary available at: http://echa.europa.eu/registration-dossier/-/registered-dossier/15201/6/2/2/?documentUUID=85b4b3f7-7374-4e44-a378-c87738dd654e. Accessed 13 May 2015

Rufli H, Mueller K (1985d) Report on the test for acute toxicity of TK 10504 to rainbow trout. OECD guideline no. 203 Paris (1984). Ciba-Geigy Ltd, Basle, Switzerland, Robust study summary available at: http://echa.europa.eu/registration-dossier/-/registered-dossier/15201/6/2/2/?documentUUID=2713bfaa-a835-45f5-b0cc-be6e96fb6571. Accessed 13 May 2015

Salinas E (2011) 4,4′-MDA: acute toxicity to *Daphnia magna*. International Isocyanate Institute, Manchester, UK, III Report no. 11612. Robust study summary available at: http://echa.europa.eu/registration-dossier/-/registered-dossier/15201/6/2/4/?documentUUID=5d6b9a75-0832-40fa-bf01-37ff9c025164. Accessed 13 May 2015

Salinas E (2012) Oligomeric MDA: acute toxicity to *Daphnia magna*. International Isocyanate Institute, Manchester, UK, III Report No. 11624. Robust study summary available at: http://echa.europa.eu/registration-dossier/-/registered-dossier/14114/6/2/4. Accessed 13 May 2015

Schaefer EC, Ponizovsky A (2013) 4,4′-MDA: aerobic and anaerobic transformation in aquatic sediment systems. International Isocyanate Institute, Manchester, UK, III Report No. 11646. Robust study summary available at: http://apps.echa.europa.eu/registered/data/dossiers/DISS-

9c7b34e3-c5c4-7414-e044-00144f67d249/AGGR-acfea30d-ad97-4274-b8ce-aabd18d556da_DISS-9c7b34e3-c5c4-7414-e044-00144f67d249.html. Accessed 3 March 2016

Schwarz H (2013) 4,4′-MDA: effect on nitrification in soil. International Isocyanate Institute, Manchester, UK, III Report No. 11643. Robust study summary available at: http://echa.europa.eu/registration-dossier/-/registered-dossier/15201/6/4/5. Accessed 13 May 2015

Schupp T (2014) Methylene-4,4′-dianiline: odd in nitrification inhibition tests. Society of Environmental Toxicology and Chemistry—German Language Branch (SETAC-GLB), September 9–11, Giessen, Germany, p 13. http://www.setac-glb.de/GLB-Tagungen-1996-2014.266.0.html. Accessed 8 May 2015

Tatsumi K, Freyer A, Minard RD, Bollag JM (1994) Enzyme-mediated coupling of 3,4-dichloroaniline to ferulic acid: a model for pollutant binding and humic materials. Environ Sci Technol 28:210–215

Tullner WW (1960) Endocrine effects of methylenedianiline in the rat, rabbit and dog. Endocrinology 66:470–474

Van der Hoeven N, Roza P, Henzen L (1992a) Determination of the effect of TDI, TDA, MDI and MDA on the emergence and growth of the plant species *Avena sativa* and *Lactuca sativa* according to OECD guideline no. 208. International Isocyanate Institute, Manchester, UK, III Report No. 11024. Robust study summary available at: http://echa.europa.eu/registration-dossier/-/registered-dossier/15201/6/4/4/?documentUUID=2f3a4ca6-39ba-4764-9977-f24dd99dedc5. Accessed 13 May 2015

Van der Hoeven N, Roza P, Henzen L (1992b) Determination of the LC50 (14 days) of TDI, TDA, MDI and MDA to the earthworm *Eisena fetida* according to OECD guideline no. 207. International Isocyanate Institute, Manchester, UK, III Report No. 11025. Robust study summary available at: http://echa.europa.eu/registration-dossier/-/registered-dossier/15201/6/4/2. Accessed 13 May 2015

Weber EJ, Spidle DL, Thorn KA (1996) Covalent binding of aniline to humic substances. 1. Kinetic studies. Environ Sci Technol 30:2755–2763

West RJ, Henry KS, Davis JW (2004) Preliminary study to determine appropriate conditions for evaluating chronic toxicity of TDA to sediment-dwelling organisms. International Isocyanate Institute, Manchester, UK, III Report No. 11495

Whiteman FW, Ankley GT, Kahl MD, Rau DM, Balcer MD (1996) Evaluation of interstitial water as a route of exposure for ammonia in sediment tests with benthic macroinvertebrates. Environ Toxicol Chem 15(5):794–801

Zhang L-L, Wu Z-J, Shi Y-F, Chen L-J, Song Y-C, Juan Y-H (2010) Inhibitory effects of aromatic compounds on soil nitrification. Pedosphere 20(3):326–333

# Cadmium Bioavailability, Uptake, Toxicity and Detoxification in Soil-Plant System

**Muhammad Shahid, Camille Dumat, Sana Khalid, Nabeel Khan Niazi, and Paula M.C. Antunes**

## Contents



M. Shahid (✉) • S. Khalid
Department of Environmental Sciences, COMSATS Institute of Information Technology, Vehari 61100, Pakistan
e-mail: shahidzeeshan@gmail.com; sanakhalid@ciitvehari.edu.pk

C. Dumat
Centre d'Etude et de Recherche Travail Organisation Pouvoir (CERTOP), UMR5044, Université J. Jaurès—Toulouse II, 5 Allée Antonio Machado, 31058 Toulouse Cedex 9, France
e-mail: camille.dumat@ensat.fr

N.K. Niazi
Institute of Soil and Environmental Sciences, University of Agriculture Faisalabad, Faisalabad 38040, Pakistan

Southern Cross GeoScience, Southern Cross University, Lismore 2480, NSW, Australia
e-mail: nabeelkniazi@gmail.com

P.M.C. Antunes
AquaTox Testing & Consulting Inc., Guelph, ON, Canada
e-mail: pcypas@gmail.com

 
P. de Voogt (ed.), *Reviews of Environmental Contamination and Toxicology, Volume 241*, Reviews of Environmental Contamination and Toxicology 241,
DOI 10.1007/398_2016_8

# 1 Introduction

Cadmium (Cd) accumulation in plants and the subsequent trophic transfer along the food chain poses a serious threat to both ecosystem and human health (Zhang et al. 2014; Goix et al. 2014). Human exposure to Cd in excessive concentrations has the potential to cause cancer, bone lesions, lung insufficiency, cancer, teratogenic effects, renal disturbances, anemia, hypertension and weight loss (Gad 2014). Because of it's potential to cause such adverse effects, Cd is classed as potential human carcinogen (group 2B) by the US Environmental Protection Agency (EPA), a human carcinogen (group 1) by the International Agency for Research on Cancer of the World Health Organization (WHO), and ranked No. 7 among the top 20 priority hazardous substances by the Agency for Toxic Substances and Disease Registry (ATSDR 2012). The main rout of Cd exposure to humans is through the diet (Chaney 2012). This is well exemplified by the *Itaï–Itaï* (ouch ouch) disease outbreak in Japan, which caused softening of the bones and kidney failure in those affected from the consumption of Cd contaminated rice (Li et al. 2011). Because of the potential health risks, Cd contamination in rice has been of mounting concern in China (Fang et al. 2014). The use of Cd has been restricted in many European countries owing to its high potential public health impacts.

Although Cd is extremely rare in the Earth's crust, various anthropogenic activities (e.g., Zn mining and smelting, application of sewage sludge/compost as well as fertilizer, pesticide and insecticide applications) (Rao and Kashifuddin 2012) have resulted in elevated Cd-soil concentrations. Cadmium is easily taken up by plants because of its high mobility within the soil–plant system (Nedjimi and Daoud 2009; Gill et al. 2012). Cadmium phyto-uptake is affected by a number of factors related to soil properties (soil particle size, pH, temperature, cation exchange capacity) and plant physiology (root surface area, rate of root exudation and transpiration) (Rosén et al. 2012). The majority of plant species accumulate Cd in root tissues, with only very small portions being transferred to the shoots (Podazza et al. 2012). However, high Cd levels in aerial plant parts are observed in Cd hyperaccumulators (Douchiche et al. 2012; Huguet et al. 2012).

Excessive Cd accumulation in plant tissue has the potential to result in numerous morphological, physiological, and biochemical toxic effects. Cadmium is not known to participate in any vital process in living organisms. Because of its capacity to elicit adverse effects, several literature reviews have been devoted to different aspects of Cd phytotoxicity (Page et al. 1987; Holmgren et al. 1993; Kabata-Pendias 1993; Toppi et al. 2012; Gallego et al. 2012). Toxicity of Cd to plants is generally associated with disruption of water and nutrient uptake and transport (Kabata-Pendias 1993; Zhang et al. 2010; Douchiche et al. 2012), alteration of nitrogen metabolism (Dguimi et al. 2009), disruption of ATPase activity (Lin et al. 2012), reduced photosynthesis, plant growth and respiration, dysfunction of the plant photosynthetic machinery in chloroplasts and stomatal closure (Hasan et al. 2011). Other symptoms of Cd toxicity may include reduced plant length and browning of roots, leaf rolling, chlorosis, and necrosis (Najeeb et al. 2011).

At the cellular level, excessive Cd concentrations can cause chromosomal aberrations and alteration of cell cycles and division (Inglot et al. 2012; Filipič 2012) as well as enhanced production of reactive oxygen species (ROS) (Lin et al. 2012; Gill et al. 2013). Overproduction of ROS may cause cell death due to oxidative processes such as DNA and RNA damage, enzyme inhibition, protein oxidation and membrane lipid peroxidation (Tian et al. 2011; Shahid et al. 2013a, 2014a). To minimize Cd-induced ROS formation, plants possess several defense strategies, including sequestration of the metal within vacuoles and chelation by organic molecules (Hossain et al. 2012). As a secondary defense mechanism, plants can also activate antioxidants to fight against the overproduction of Cd-induced ROS (Chen et al. 2012; El-Boshy et al. 2014). Several genes have been suggested to play a role in Cd detoxification and/or transport (Fässler et al. 2011).

The basic concepts of Cd bioavailability, uptake, toxicity and detoxification in soil-plant system have been well documented in studies carried out and published before 2000. These earlier findings/facts have been summarized in several books and review articles (e.g., Page et al. 1987; Holmgren et al. 1993; Kabata-Pendias 1993; Das et al. 1997). During the last decade, several articles have additionally been produced which focus on the biogeochemical behavior of Cd in the soil-plant system, highlighting the mechanisms involved at the cellular level. The present review summarizes the latest data (published after 2000) on Cd bioavailability in soil, plant uptake and translocation, toxicity, and detoxification – aspects which are highly relevant to soil remediation and risk assessment. The review is divided into eight sections: (1) introduction; (2) global use of Cd; (3) behavior of Cd in soil; (4) soil-plant transfer of Cd; (5) Cd phytoxicity; (6) Cd detoxification mechanisms; (7) hermetic effects, and; (8) conclusions and perspectives.

## 2 Global Use of Cadmium

Cadmium is a naturally occurring element in the Earth's crust and waters. Cadmium ore is not mined for production because more than enough is produced as a byproduct from mining, smelting, and refining of zinc sulphide ore (Rao et al. 2010). For this reason, global Cd production depends more on Zn refining than marketplace demand. The proportion of Cd and Zn varies from mine to mine, ranging from 0.07 % to 0.83 % with an average value of 0.23 % (OECD 1994). Recycling of steel and iron scrap is the secondary source for Cd production (about 10–15 % of consumption) (OECD 1994).

Despite its numerous industrial, agricultural and domestic uses, Cd was not widely utilized until World War-I (Rao and Kashifuddin 2012). Owing to its useful properties, Cd is now used in several industrial activities. Today, the major global consumption of Cd (almost 80 %) can be attributed to its use in rechargeable nickel–Cd batteries (USGS 2015). However, nickel–Cd batteries are gradually being phased out and replaced by nickel metal hydride batteries. Cadmium was often used in the steel industry for protecting steel from corrosion by Cd electroplating. Cadmium also has a number of other uses, including the preparation of alloys (due to its good fatigue resistance and a low coefficient of friction) and as a stabilizer for plastics (Scoullos et al. 2001; Nzengue et al. 2015). Cadmium telluride thin-film photovoltaics are used as a substitute to the traditional silicon-based solar cells (USGS 2015), and are preferred for large-scale, ground-mounted utility systems and commercial rooftop applications (USGS 2015).

Between 2001 and 2003, Cd production decreased from 18,700 to 15,000 thousand metric tons, but increased to 18,800 thousand metric tons in 2004. After 2004, production of recoverable Cd increased gradually, reaching 22,200 thousand metric tons in 2014 (USGS 2015; Fig. 1), with China, Korea and Japan being the major producers (USGS 2015). Given that Cd-production in mining increased overall by 15 % between 2001 and 2014 alone (Fig. 1), it is very important to understand the biogeochemical behavior of Cd, and the impact that increased Cd concentrations will have on the environment.

## 3 Behavior of Cd in Soil

### *3.1 Cd Level and Sources in Soil*

Cadmium was first discovered in Germany simultaneously by Stromeyer and Hermann in 1817 as a by-product of the Zn refining process. Approximately 700 times less abundant than Zn, Cd ranks as the 65th most abundant element in the Earth's crust (Emsley 2011), with a lithosphere concentration of approximately 0.1–0.2 mg/kg (Heinrichs et al. 1980). While Cd can be found together with the Zn ore spalerite, greenockite (CdS) is the only known Cd-based ore (Rao et al. 2010).

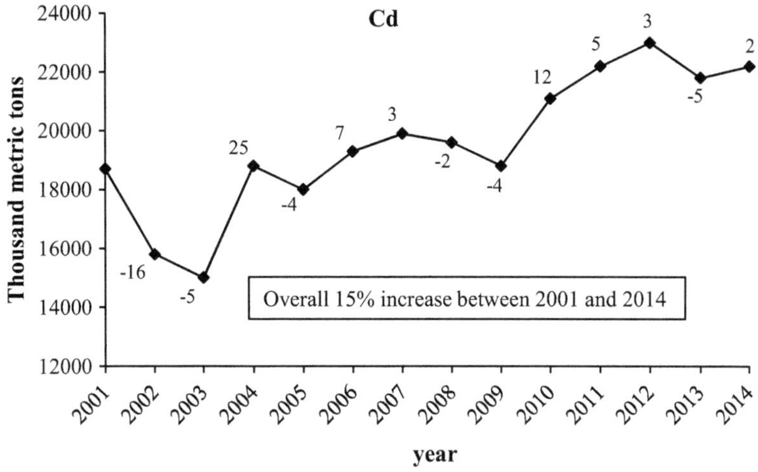


**Fig. 1** Annual world mine production of Cd in thousand metric tons (source, USGS 2015). The *plus and minus values/labels* in the graph line indicate the % increase and decrease of Cd production. Despite restriction on the use of Cd in many European countries owing to its high potential public health impacts, the production of Cd has increased by 17 % between 2001 and 2013

The natural concentration of Cd in soil is between 0.07 to 1.1 mg/kg in many parts of the world (Thornton 1992). While natural concentrations vary with soil type, higher Cd concentrations can be expected for soils overlying sedimentary rocks (which contain the highest Cd concentrations ranging between 0.3 to $\geq$15 mg/kg) (Traina 1999), and soils impacted by Cd-phosphate fertilizers (Liu et al. 2012). Except for soil and coal, concentrations of Cd for different constituents of the Earth's crust (Table 1) generally do not exceed 0.6 mg/kg. Cadmium concentrations higher than 10 mg Cd/kg soil can be toxic to plants (Holmgren et al. 1993).

According to an extensive survey carried out in the United States with 3305 soil samples collected from agricultural areas in 36 states, Cd soil levels varied from 0.005 to 2.4 mg/kg, with mean values of 0.27 mg/kg (Holmgren et al. 1993). According to FOREGS, total Cd concentrations in European agricultural soil range between 0.06 to 0.6 mg/kg (Salminen et al. 2005). Similarly, concentrations ranged from $<$ 0.1 to 2.9 mg Cd/kg for surface soils collected from 237 farms in Ohio (Logan and Miller 1983), and 0.06 to 0.74 mg/kg for 16 soil series in Minnesota (Pierce et al. 1982). Burau et al. (1973) reported that Cd concentrations ranged from 0.05 to 10.1 mg/kg for 177 agricultural soils developed from Monterey shale deposits in Salinas Valley, California. The variation in Cd concentrations among different soils reflects in part the concentration differences in the parent material from which these soils are formed. To minimize the risk of elevated Cd concentrations in edible produce and Cd-induced phytotoxicity, it is recommended that soils containing more than 1 mg Cd/kg dry soil not be used for agricultural or horticultural purposes (Louwagie et al. 2009), and that remedial action be taken for

**Table 1** Cadmium ranges/level in different constituents of Earth's crust

| Constituents of Earth's crust | Cd level (mg/kg) | References |
|---|---|---|
| Lithosphere | 0.1–0.2 | Heinrichs et al. (1980) |
| Ultramafic rocks | 0.03–0.05 | Mesjasz-Przybyłowics et al. (2004) |
| Mafic rocks | 0.18 | Ivanov (1996) |
| Intermediate rocks | 0.13 | Ivanov (1996) |
| Clay, clay shales | 0.3 | Ivanov (1996) |
| Sandstones | 0.01–0.41 | Page et al. (1987) |
| Limestones | 0.001–0.5 | Page et al. (1987) |
| Granite | 0.001–0.6 | Page et al. (1987) |
| Basalt | 0.006–0.6 | Page et al. (1987) |
| Coal | 1.0 | Tauber (1988) |
| Subsoil | 0.09 | FOREGS |
| Topsoil | 0.15 | FOREGS |
| Soil | 0.07–1.1 | Thornton (1992) |

soils containing 5–20 mg/kg Cd (Eikmann et al. 1993). Unfortunatley, several previous studies have shown high Cd contamination in agricultural soils where food has been grown (Zhao et al. 2015).

Volcanoes and weathering of rocks are the major natural sources of Cd mobilisation from the Earth's crust (Choppala et al. 2014; Zhao et al. 2015). Collectively, natural sources are estimated to release about 150–2600 tonnes of Cd to air (Nordic Council of Ministers 2003). In 1983, about 140–1500 tonnes of Cd was emitted to the atmosphere from volcanoes (Nriagu 1989).

Relative to natural processes such as volcanic activity and the weathering of rocks, anthropogenic activities such as mining, combustion of fossil fuels, metal production and other industrial and agricultural practices have been responsible for the increased Cd levels in the environment (Choppala et al. 2014; Agnieszka et al. 2014; Zhao et al. 2015). Industrial emission, animal manures, sewage sludge and phosphate fertilizers are reported to have increased Cd concentrations in agricultural soils (Rao and Kashifuddin 2012). Munshower (1977), for example, reported high Cd levels (29 mg/kg) in Montana up to two km northeast of a Pb-Zn smelter, which decreased with distance and reached near background levels after 24 km. Recently, Shahid et al. (2013b) reported high levels of Cd (70.2 mg/kg) within a 40 $km^2$ area surrounding the Metaleurop smelter near Evin-Malmaison (North of France).

Although atmospheric deposition has decreased during the last two to three decades in developed countries, it remains a major source of Cd release (e.g., 2500–15,000 tonnes of Cd annually) to agricultural soils (Shahid et al. 2013b; UNEP 2010). The use of wastewater for crop irrigation without any prior treatment in peri-urban areas is another major source of Cd for agricultural soil. Singh et al. (2010) reported 0.02 mg/l of Cd in wastewater of Varanasi, India. Despite low levels of Cd in wastewater, sorption of Cd by the soil may result in Cd accumulation over time. Soil amendments such as municipal sewage sludge (MSS) may also contain high levels of Cd (e.g., 2–3500 mg Cd/kg sludge) (Logan and Chaney 1984; Page et al. 1987). It is estimated that the application of MSS @ 5 tons/ha, for MSS containing 10–15 mg/kg Cd, can lead to an increase in the surface soil Cd concentration of 10–15 %. Landfills and various deposits of discarded products further release about 7500–29,500 tonnes of Cd per year to soil (UNEP 2010), as do agrochemicals used for enhanced crop production (e.g., 8.9 g $ha^{-1}$ $year^{-1}$ from phosphorus fertilizers; Bramley 1990). Among the fertilizers, those that are phosphate-based and manufactured from seabed sediments with high Cd content have the potential to contain the highest Cd concentrations (Louwagie et al. 2009). In various countries of Europe (Denmark, Ireland, Austria, Finland, Greece and the United Kingdom), Cd is estimated to have increased by 7–43 % as a result of the past 100 years of fertilizer application (UNEP 2010).

## 3.2 *Bioavailability of Cd in Soil*

In order to estimate the ecological risk associated with Cd contamination in soil, it is necessary to understand the fraction of total Cd available for uptake by biota. Total Cd concentrations in soil do not necessarily reflect Cd phytoavailability (Duplay et al. 2014) because different chemical forms (chemical speciation) and the varying distribution of Cd (adsorbed or free) affect its phytoavailability. Cadmium is relatively available for plant uptake because it is predominantly found in soil bound to the exchangeable solid phases (Degryse et al. 2012) and thus readily released into the soil solution. In the soil solution, Cd is mainly present as the $Cd^{2+}$ ion, or as inorganic or organic complexes. In the solid phase, Cd may be reversibly bound to soil particulates, such as organic matter or Fe and Mn oxides. Plant uptake of Cd occurs mainly through contact with porewater, which is the net result of Cd partitioning between solid and liquid phases of soil. However, recent studies have also reported metal uptake by plants via foliar transfer (Uzu et al. 2010; Schreck et al. 2012a, b). According to Xiong et al. (2014a), foliar uptake of Cd and Pb due to particulate matter deposition is of particular importance near recycling factories.

The chemical forms and partitioning of Cd in soil is governed by a number of reactions that involve precipitation/dissolution, adsorption/desorption, and Cd–ligand complex formation (Bolan et al. 2013). These dynamic processes are strongly affected by variations in redox conditions (Zhang et al. 2012), cation exchange capacity (Jiang et al. 2012), soil pH (Saeki and Kunito 2012), soil texture (Qi et al. 2012), biological and microbial conditions (Joubert et al. 2007), metal burdens (Gao et al. 2009), organic and inorganic ligands (Niazi et al. 2011a; Shahid et al. 2014b; Austruy et al. 2014), competing cations (Helios-Rybicka and Wójcik 2012) and temperature (Silber et al. 2012) (Fig. 2).

The chemical speciation of Cd, and its partitioning in soil, is a key factor in controlling both the fate and toxicity of Cd in the soil-plant system (Pizzol et al. 2012; Landrot et al. 2012). It is well-known that total soil Cd concentrations

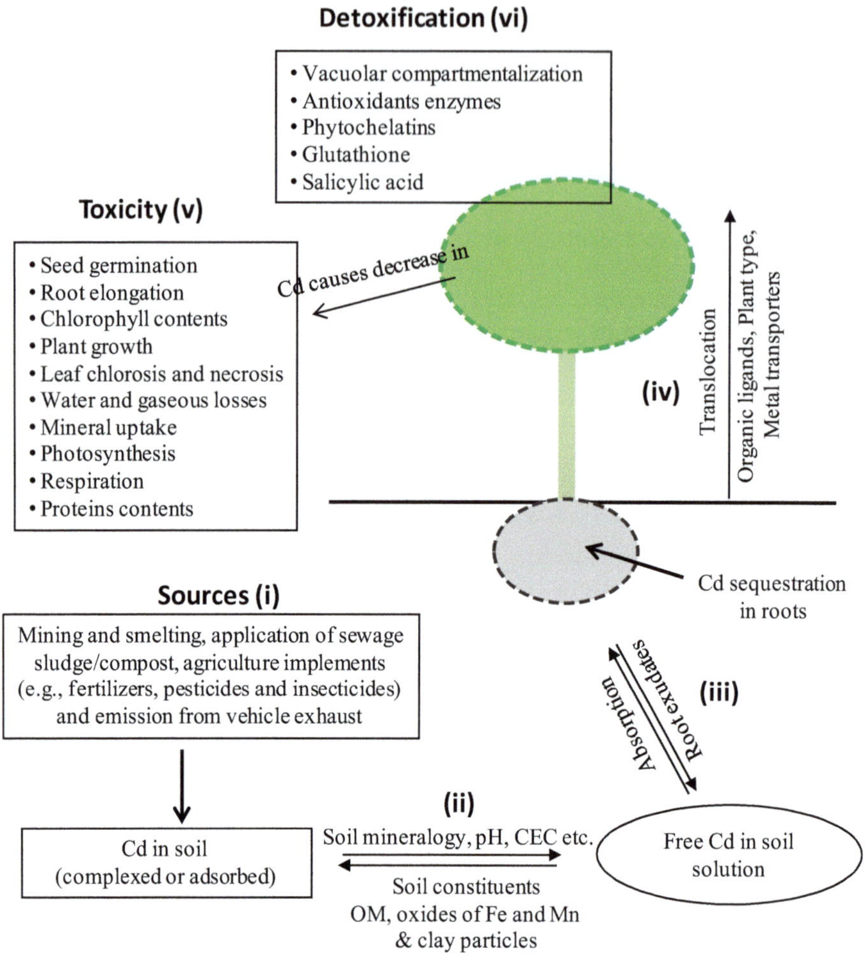


**Fig. 2** Biogeochemical behavior of Cd in soil-plant system. (i) Accumulation of Cd in soil from different natural and anthropogenic sources, (ii) effect of soil physic-chemical propertie on the complexation, mobility and bioavailability of Cd in soil, (iii) Cd uptake by plants via root transfer and major sequestration (>90 %) in root tissues, (iv) translocation to aerial parts depending on the nature of plant and applied Cd level and (v) toxic effects of Cd to plants and (VI) detoxification mechanisms involving the role of antioxidants, phytochelatins, glutathione and salicylic acid

tend to be a poor proxy for predicting Cd mobility, solubility, extractability, and bioavailability (Udovic and McBride 2012). This is because the total sum of all chemical forms of Cd are rarely 100 % bioavailable (Udovic and McBride 2012), and the biogeochemical behavior of Cd depends largely on the free Cd concentration (Dabrin et al. 2012). For these reasons, knowledge of Cd speciation is considered necessary for informing soil remediation and risk assessment studies (Niazi et al. 2011b; Shahid et al. 2011, 2012a; Bade et al. 2012).

Traditionally, analytical techniques used to assess metal speciation in soil and water have included anodic stripping voltammetry, ion selective electrodes, competitive ligand equilibration/adsorptive stripping cathodic voltammetry, and sorption onto $C_{18}$ columns (Cobelo-García et al. 2005). However, these techniques are very complex and time consuming. When actual measurements are not feasible, computer speciation models (capable of performing complex calculations) offer a cost-effective alternative (Shahid et al. 2012b, 2014b), although site-specific validation is required to confirm the accuracy of the model's output. The most widely used speciation models include: the Windermere Humic Aqueous Model VI (WHAM VI), CHESS, CHEAQS, PHREEQC and VISUAL MINTEQ (Tipping et al. 1998; Parkhurst and Appelo 1999; Gustafsson, 2008; Antunes and Kreager 2014; Shahid et al. 2014d). These models offer the advantage of desk-top computing and are appropriate for both small and large scale applications. However, the accuracy of the estimated speciation may vary among models and chosen input values (e.g., Cd concentration range, DOC concentration and quality, consideration of precipitation reactions).

## 3.3 *Effect of Soil Chemical Properties on Cd Bioavailability*

### 3.3.1 Effect of Soil pH on Cd Bioavailability

Soil pH is one of the most important parameters governing Cd speciation, partitioning, and bioavailability (Ardestani and van Gestel 2013), as evidenced by the negative correlation established between soil pH and Cd mobility/phytoavailability (Bilodeau-Gauthier et al. 2011). Cadmium is reported to exist in different chemical forms under different soil pH conditions. In most soils, 99 % of the total Cd is associated with soil colloids. Cadmium may occur as cationic species: $CdHS^+$, $CdOH^+$, $CdHCO_3{}^+$, $CdCl^+$, and as anionic species: $Cd(HS)_4{}^{2-}$, $Cd(OH)_3{}^-$, $Cd(OH)_4{}^{2-}$, $CdCl_3{}^-$ (Kabata-Pendias and Sadurski 2004). Of the Cd in soil solution, it has been estimated that between 55 and 99 % exists in free ionic form (Kabata-Pendias 1993). At low pH, Cd in soil solution is predominately present as $Cd^{2+}$, $CdSO_4$ or $CdCl^+$, whereas in alkaline solutions the less bioavailable $CdHCO_3{}^+$, $CdCO_3$ or $CdSO_4$ species predominate (Sauvé et al. 2000). Cadmium solubility greatly depends on the acidity of the soil solution. For example in mineral soils having a pH range of 4.0–4.5, a decrease in pH of merely 0.2 units results in a 3–5 times increase in the labile Cd pool. This is in contrast to an increase in pH within the basic pH range, which results in a greater extent of Cd adsorption on to soil particles with a subsequent reduction in plant uptake of Cd, Zn, and Pb (Kuo et al. 1984). Douay et al. (2009) and Fritsch et al. (2010) reported that the spatial distribution of extractable and total Cd levels in 262 soil samples over an area of 40 $km^2$ in Northern France greatly varied in accordance with soil pHand other soil properties. Because of the relationship between pH and Cd bioavailability, one strategy used for Cd phytoremediation has been to lower soil pH to enhance plant Cd uptake (Shahid et al. 2011). In contrast, for both plant and groundwater protection, increasing soil pH is likely to be an effective measure to reduce phytoavailability, since Cd precipitation and adsorption on soil particles should minimize soil solution Cd concentrations (Sauvé et al. 2000; Qi et al. 2012). Although impacts of soil pH on the formation of Cd precipitates vary with soil type, high redox potential and low soil pH have generally been shown to restrict the formation of Cd-precipitates (Li et al. 2011). In acidic soils, Cd solubility is mainly controlled by soil organic matter and oxides and hydroxides of Fe and Mn (Jiao et al. 2012). Above pH 7.5, sorbed and precipitated Cd (e.g., as $CdCO_3$, and possibly $Cd_3(PO_4)_2$) is not readily mobilized (Kabata-Pendias 2011). In calcareous soils where Cd-carbonate complexes predominate (Renella et al. 2004), a large proportion (e.g., 35 %) of Cd may additionally be present as $Cd^{2+}$ and $CdHCO_3^+$ (Hirsch and Banin, 1990).

Under conditions of low soil pH, Cd bioavailability and mobility is higher due to Cd transformation from immobile forms such as Fe and Mn oxides and carbonates to more bioavailable and exchangeable forms (Xian and Shokohifard 1989; Li et al. 2014). Soil pH may decrease or increase Cd bioavailability in soil indirectly via transformation of Fe and Mn minerals. Iron oxides have a significant capacity to sorb and immobilize Cd (Liu et al. 2014a, b). Thus, secondary Fe minerals formed by Fe(III) reducing bacteria and the Fe bacteriogenic oxides formed by Fe(II)-oxidizing microorganisms can reduce Cd bioavailability (Cortés-Martínez et al. 2004; Muehe et al. 2013), whereas the reduction of Cd bearing Fe(III) minerals by Fe(III)-reducing bacteria can increase Cd bioavailability (Muehe et al. 2013). Different mechanisms have been proposed for the negative correlation between soil pH and Cd dissolution/desorption. Soil pH governs the degree of net charge associated with the solid-phase: the higher the pH the more negatively charged the solid-phase surface and vice versa. Under acidic soil conditions, $H^+$ ions compete with Cd for binding sites, which results in Cd desorption from soil particles into solution. In contrast, under alkaline conditions, the solid-phase exchange sites are readily available for metal cation binding. Additionally, at high pH levels, Cd is adsorbed more strongly by the solid phase due to hydrolysis of Cd to hydroxy species (Barrow 1986) and higher specific adsorption of Cd (He et al. 2001), thus, making metal desorption more difficult at high pH (Tudoreanu and Phillips 2004).

### 3.3.2 Effect of Soil Organic Matter on Cd Bioavailability

Soil organic matter, originating from the decomposition of microbial, plant and animal material, plays an important role in governing Cd mobility/bioavailability (Quenea et al. 2009) as it readily forms complexes with Cd. The effect of humic substances (HSs) on metal bioavailability varies with its form in soil (i.e., either suspended or dissolved), as well as with the concentration, source, and physico-chemical quality (e.g., molar mass, humification stage, types of functional groups) (Cabaniss et al. 2000; Calace and Petronio 2004). Humic substances have been reported to bind $Cd^{2+}$ to a greater extent than the major inorganic ligands, especially at high pH values (Reuter and Perdue 1977). Compared to the humic acid component of HSs, fulvic acids are more reactive due to their smaller size, higher oxygen content, and higher cation exchange capacity (Zeng et al. 2002; Shahid et al. 2012d). Soils with a higher organic matter content are more effectively able to reduce Cd uptake by plants as a result of Cd-sorption (onto carboxylic and phenolic hydroxyl groups (Dumat et al. 2006; Shahid et al. 2012d)) and subsequent removal of Cd from the soil solution.

Since dissolved HSs form soluble complexes with Cd, soil amendments with OM may also enhance Cd mobility and uptake by plants (Almås and Singh 2001). This may in part be why some authors have reported a decrease in metal bioavailability in the presence of HS (Guerrero et al. 2000), while others have shown an increase (Sánchez-Marín et al. 2010). In addition to these direct impacts, OM can also influence Cd bioavailability by altering the soils' physico-chemical properties such as CEC, pH, particle size distribution, porosity and cracking pattern, microbial and enzyme activities, and soil solution composition (Turrión et al. 2012).

### *3.4 Influence of Soil Microbial Activity on Cd Behavior in Soil*

Soil microbial activity is known to increase Cd availability via excretion of organic acids and subsequent solubilization of Cd-bearing minerals (Wang et al. 2009; Liu et al. 2014a, b; Ahmad et al. 2015). Soil amendment with Cd-solubilizing microorganisms such as plant growth-promoting bacteria (PGPR) is a very effective technique for enhancing Cd bioavailability (Belimov et al. 2005). Xiong et al. (2008) for example, observed significant increases in the uptake of Cd, Zn, and Pb by *Sedum alfredii* Hance when soils were inoculated with rhizosphere microorganisms. Similarly, the capacity of *Thlaspi caerulescens* to extract Cd, Pb and Zn has been found to increase following microbial inoculation of the soil (Epelde et al. 2010). This finding supports the work of Duponnois et al. (2006), who showed that inoculation of *Sorghum bicolor* with Cd-tolerant bacteria significantly increased Cd uptake, and also the work of Sheng and Xia (2006), who observed that soil inoculation with isolates containing bacteria caused a 74 % increase in the Cd content of *Brassica napus*. Arbuscular mycorrhizal fungi appear to play a similar role, as suggested by the results of Long et al. (2010), who found that inoculation of *Perilla frutescens*, *Litsea cubeba*, *Dysphania ambrosioides*, *Phytolacca Americana* and *Rehmannia glutinosa* resulted in enhanced uptake of Cd, Zn, Cu and Pb.

In contrast, some soil-inoculated microorganisms, such as *Azospirillum lipferum*, *Arthrobacter mysorens*, *Agrobacterium radiobacter* (Belimov et al. 2004), and arbuscular mycorrhizae (Karagiannidis and Nikolaou 2000) have been shown to decrease Cd phytoavailability. Recently, Dary et al. (2010) reported a decrease in the uptake of Cu, Cd and Pb by *Lupinus luteus* when inoculated with *Brady rhizobium*. The inoculation of Cd-polluted soil with *Ralstonia eutropha* has also been shown to reduce Cd uptake by *Nicotiana bentaminana* as a result of Cd immobilization (Valls et al. 2000). Tiwari et al. (2008) found that 11 bacterial strains of *Typha latifolia* reduced Cd bioavailability in fly ash. Similarly, *Pseudomonas aeruginosa* was also reported to decrease Cd uptake by >50 % in *Cucurbita maxima* in Cd-contaminated soil (Sinha and Mukherjee 2008). The decrease in bioavailability of Cd by soil microorganisms might be due to Cd accumulation by the microbes themselves, or Cd immobilization in the soil. The latter hypothesis is supported by the work of Xu et al. (2012), who found that bacteria transformed 8–25 % of Cd to a less bioavailable organic-bound fraction.

# 4 Soil-Plant Transfer of Cd

## *4.1 Molecular Understanding of Cd Uptake by Plants*

Although Cd is a non-essential element, its phytoaccumulation factor index (ratio of Cd level in plant to soil) is high, and may surpass that of many essential nutrients (Gonçalves et al. 2009). Cadmium does not possess any specific cell entry pathway. Cadmium enters the plants from soils mainly via root uptake in the form of accidental transport by specific and non-specific transporters of essential elements (Llugany et al. 2012), such as $Fe^{2+}$, $Ca^{2+}$, $Zn^{2+}$, $Cu^{2+}$ and $Mg^{2+}$ (Roth et al. 2006; Mendoza-Cózatl et al. 2011). To date, no Cd-specific transporters have been unambiguously identified.

Transcriptional and post-transcriptional gene regulations responsible for encoding specific and non-specific transporters are complex. ZIP (ZRT, IRT-like protein) family transporter for $Fe^{2+}$, $Zn^{2+}$ and $Mn^{2+}$ have been reported to mediate Cd uptake in plant roots (Mendoza-Cózatl et al. 2011). It has been demonstrated that Cd enters root cells via ZIP transporters or through cation channels (Verbruggen et al. 2009; Gallego et al. 2012). OsIRTs is also suggested to contribute to Cd uptake by plants (Nakanishi et al. 2006; Lee and An 2009). The role of OsIRTs in Cd uptake varies with plant type. For example, OsIRTs has been found to have no effect on Cd uptake by Prayon ecotype (*T. caerulescens*), which is less capable of accumulating Cd (Lombi et al. 2002). Cadmium can also enter plant root cells by Yellow-Stripe 1-Like proteins as Cd-chelates (Curie et al. 2009) and NRAMP (natural resistance-associated macrophage protein) family transporters. Similarly, TcZNT1 has been shown to mediate low-affinity Cd uptake in *T. caerulescens* (Pence et al. 2000), and a low affinity cation transporter (LCT1) cDNA of *Triticum aestivum* has been reported to enhance Cd uptake in yeast cells (Clemens et al. 1998). It has been proposed that LCT1 functions as a Ca transport system in plants and might contribute to Cd transport (Clemens et al. 1998). Despite these advances in our understanding, the mechanisms/processes involved in Cd uptake by plants are not yet well understood.

## 4.2 Competition Between Cd and Cations for Uptake

Although further research is still needed, we know that metal-metal interactions influence Cd uptake. The presence of $La^{3+}$, $Ca^{2+}$, $Cu^{2+}$, $Fe^{2+}$, $Zn^{2+}$ or $Mn^{2+}$ in the rhizosphere solution, for example, have been shown to inhibit Cd uptake by plant roots (Lux et al. 2011). These cations likely affect Cd uptake as a result of competition for ion transport specific channels (Wojas et al. 2009). This hypothesis is supported by the work of Li et al. (2012) who showed that pre-treatment with the $Ca^{2+}$ channel blockers $LaCl_3$ and verapamil significantly reduced $Cd^{2+}$ influx into roots. Although selenium would not be expected to compete with Cd for uptake in the same manner as cationic metals, Hu et al. (2014) similarly reported reduced uptake (44 %) of Cd by *Oryza sativa* in the presence of selenium.

Ion-specific interactions of trace metals with Cd, and the subsequent impact on plant-Cd uptake, vary with plant genotype and the type and concentration of metals present in the soil solution (or rhizosphere) (Lugon-Moulin et al. 2004). Cadmium and Zn, for example, have been found to behave alike in soil- water- plant system owing to their chemical similarities (Mengel et al. 2001). The similarities between these two metals result in a protective effect of Zn against Cd uptake, whereby plants generally absorb more Cd when the soil solution concentration of Zn is low (Mengel et al. 2001). Owing to this antagonistic relationship, Cd concentrations in plant roots have been shown to decrease by up to 50 % in the presence of Zn (Lugon-Moulin et al. 2004). Thus, in the context of food chain transfer, agricultural soils or waters containing moderate concentrations of Zn are likely to afford some protection against crop contamination by Cd.

Cadmium in the rhizosphere or soil solution can interfere with the uptake of essential nutrients by plants (Castillo-Michel et al. 2009). This interaction is of public concern (Gonçalves et al. 2009), as Cd can cause severe nutrient deficiencies in plants (Dong et al. 2006). This in turn can result in physiological disorders and a diminution of growth, with specific impacts differing among plant species, cultivar, and plant tissues (Liu et al. 2003). Using energy-dispersive X-ray spectroscopy Vaculík et al. (2012) revealed that Cd disturbed the distribution of essential elements in plant tissues. They reported that Ca, Mg, Si, Na and Zn accumulated mainly in the peripheral bark, S and K in the phloem, and Cd in both vascular tissues. Jiang et al. (2004) reported that Cd reduced the uptake of K, P, Fe, Zn and Ca in the roots, and K, P, Cu and Ca in the shoots of *Brassica juncea*. In a study with *Brassica campestris*, Zhu et al. (2004) reported that Cd exposure resulted in an increase in the K concentration in the roots and decrease in the shoots, with the opposite trend being observed for P, Mg, B, Fe and Mn. For Zn, Ca and S, concentrations increased in both roots and shoots. Nedjimi and Daoud (2009) reported that the addition of Cd to the culture medium decreased both root and shoot concentrations of Ca and K in *Atriplex halimus*. Sun and Shen (2007) also reported a decrease in Fe, Mn, P, Mg and S concentrations under Cd stress in *Brassica oleracea*. Cherif et al. (2011) further reported a decrease in Zn uptake by

*Solanum lycopersicum* in the presence of Cd. Based on the available data, no definitive conclusion can be drawn regarding the impact of Cd on essential nutrients.

## 4.3 Cadmium Sequestration in Plant Roots

Subcellular accumulation of Cd in plants varies greatly not only among plant species but also among cultivars and genotypes of the same species (Xin et al. 2013), and the plants' Cd detoxification and tolerance mechanisms. For most plants, Cd tends to accumulate in the roots, with only a small portion being translocated to aerial parts (Ahmad et al. 2011). Generally, Cd contents in plants decrease in order of roots > stems > leaves > fruits > seeds (Benavides et al. 2005). Some species can sequester high concentrations of Cd in their roots with reduced translocation to shoots, or remobilize Cd from leaves to roots through the phloem (Mendoza-Cózatl et al. 2011). These are referred to as low Cd-accumulating plants. Recently, Zhou et al. (2015) reported 60–90 % accumulation of Cd in roots of 32 hybrid rice (*O. sativa*) cultivars—a percentage similar to that observed by Gonneau et al. (2014) (i.e., 50–80 %) for Cd accumulation in shoots of 22 populations of *Noccaea caerulescens*. The concentration of Cd in plant roots can be tenfold higher than in the shoot (Wang et al. 2009), suggesting that roots serve as an effective barrier to Cd translocation. The efficacy of the root cell walls in trapping Cd is attributed to the number of available functional groups (e.g., hydroxyl, carboxyl) that are able to form complexes with trace metals (Nishizono et al. 1987). For example in *Arabidopsis thalinana*, Cd was mainly bound to O/N groups in the cell wall and only a minor fraction was bound to S-containing groups (Isaure et al. 2006).

Cadmium accumulation in plant roots varies with plant type and rhizosphere Cd concentration. For example, more Cd accumulates in the cell-wall fraction of *Zea mays* than *Vicia faba* seedlings (Lozano-Rodriguez et al. 1997). Wu et al. (2005) showed that 51 % of the Cd existed as a soluble form and 36 % was present in cell walls in barley roots. Similarly, Cd is mainly concentrated in the apical cells of plant roots at low soil Cd concentration or in the proximal subapical region (Seregin et al. 2004). Cadmium mainly accumulates in nuclei and vacuoles, while a lesser amount of Cd may also be found in the plastids and cytoplasm of plant roots (Liu et al. 2007). Toppi et al. (2012) reported that Cd was concentrated in cell walls and intercellular spaces, as well as in vacuoles of cortical root cells of *Daucus carota*. Electro-dense aggregates seen using transmission electron microscopy have been found between the plasmalemma and cell wall in the outermost tissues of root, whilst very few are present in the vascular cylinder (Lux et al. 2011; Toppi et al. 2012).

## 4.4 Cadmium Translocation to Shoots

Cadmium movement from roots to aerial tissues occurs via transpiration-driven xylem loading, with symplastic and/or apoplastic transport (Hasan et al. 2009; Kranner and Colville 2011) of free $Cd^{2+}$, or Cd complexed with various chelates (Maestri et al. 2010; Lux et al. 2011). In the studies of Zhao et al. (2006), Cd shoot concentration decreased in response to a decrease in transpiration. The important role of transpiration in governing Cd translocation has been reiterated by Kranner and Colville 2011, who found that the seed Cd content differed for two rice cultivars having different transpiration rates. The process of Cd translocation to the shoot tissues is also known to occur very quickly, as evidenced by the considerable amount of Cd observed in xylem sap and shoot tissue of *O. sativa* only 1 h after Cd treatment with positron emitting radio isotopes (Fujimaki et al. 2010). Thus, Cd is translocated with water flowing through the vascular system to the leafy tissues, where water then evaporates and Cd is left to accumulate (Krzesłowska et al. 2010).

The proportion of absorbed Cd in the plant aerial parts varies according to plant characteristics (species, age, cultivar) (Foucault et al. 2013; Koren et al. 2013; Zhou et al. 2015; Van der Vliet et al. 2007). For instance, Cd translocation from roots to shoots was lower in *Solanum torvum* compared to *Solanum melongena* (Yamaguchi et al. 2011; Lux et al. 2011). Similarly, *Lupinus albus* has been found to accumulate less Cd in its shoots than *Lupinus angustifolius* and *L. luteus* (Egle et al. 2003). Pectin-rich layers of the collenchyma cell walls of the leaf veins have been reported to be the main storage sites of Cd in *Salix viminalis* (Vollenweider et al. 2006). In *A. halleri*, Cd accumulated mainly in leaf veins and trichomes (Isaure et al. 2006). Cadmium was mainly stored in the stem pitch and cortex of hyperaccumulating ecotype of *S. alfredii*, while in the non-accumulating ecotype, Cd was mainly localised in vascular bundles (Tian et al. 2011). Wójcik et al. (2005) reported that *N. caerulescens* accumulated Cd in vacuoles of spongy and palisade mesophyll cells.

The exact role of different chelators in Cd translocation has not yet been elucidated. Organic ligands such as citrate and malate are known to form weak complexes with Cd (attributed to their low binding affinities), especially at cytosolic pH values (Rascio and Navari-Izzo 2011). These and other chelating compounds are involved not only in metal hyperaccumulation and tolerance (Haydon and Cobbett, 2007), but also in overcoming metal diffusion limitiations which may otherwise restrict Cd uptake (Antunes and Hale, 2006). Amino acids, such as histidine and nicotinamine, similarly play key roles in Cd translocation owing to their ability to form stable complexes with Cd and other divalent cations (Rascio and Navari-Izzo 2011).

Xylem loading plays an important role in Cd accumulation within aerial plant parts. Low Cd loading into the xylem sap in the roots has been associated with subsequent low Cd transport to aerial parts (Hasan et al. 2009; Verbruggen et al. 2009; Santos-Echeandía et al. 2010). A number of important membrane

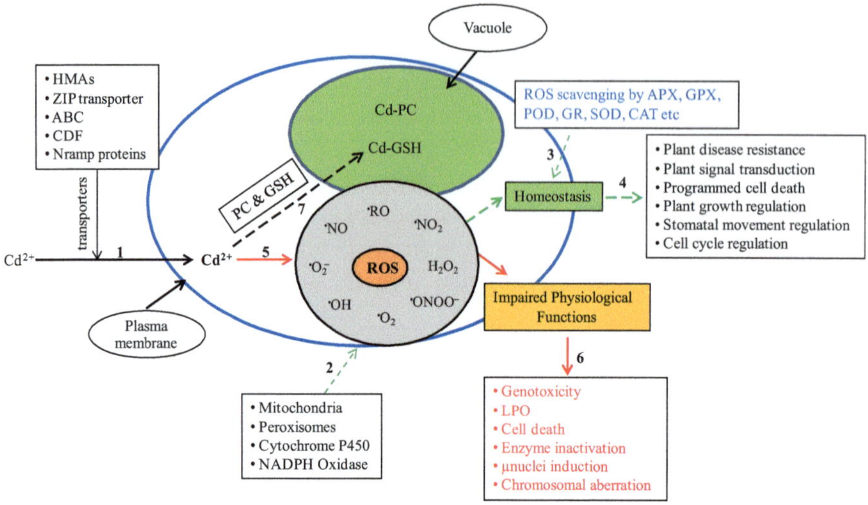


**Fig. 3** Schematic representation showing possible Cd entrance, accumulation toxicity and detoxification in plant cell. Cadmium entrance across plasma membrane is mediated by membrane transporter gene families (1). Under normal conditions, ROS are produced in different cell organellies (2) and an optimal ROS level (homeostasis) is maintained by antioxidant enzymes (3). These ROS are involved in different essential roles in living organisms (4). Presence of Cd may resulst in increased production of ROS (5), which may leads to dramatic physiological challenges to the plant termed as "oxidative stress" (6). Altrnatively, Cd may be sequestered into vacuole by PC and GSH (7). *Green arrows* indicate toxic effects of Cd whereas *green arrows* indicate Cd detoxification. HMAs, heavy metal ATPases; ZIP, ZRT and IRT-like protein; ABC, ATP binding cassette: CDF, cation diffusion facilitator; NRAMP, natural resistance associated macrophage protein; PCs, phytochelatins; GSH, glutathione; ROS, reactive oxygen species; APX, ascorbate peroxidase; GPX, guaiacol peroxidase; POD, Peroxidase; GR, glutathione reductase; SOD, superoxide dismutase; CAT, catalase; LPO, lipid peroxidation

transporter gene families, identified and characterized in recent years, are thought to be involved in Cd loading into the xylem sap and Cd transport. These gene families include the ATP binding cassette (ABC) superfamily, HMA (heavy metal ATPase), ZIP (ZRT, IRT-like protein), NRAMP (natural resistance-associated macrophage protein), YSL (yellow-stripe-like transporter), NAS (nicotinamine synthase), SAMS (S-adenosyl-methionine synthetase), FER (ferritin Fe (III) binding), CDF (cation diffusion facilitator), NRT (nitrate transporter) and IREG (iron regulated transporter) (Memon and Schröder 2009; Maestri et al. 2010) (Fig. 3).

Similarly, several studies have reported the involvement of P1B-ATPase and also ZIP family transporters in Cd transfer across the plasmalemma into the shoots (Hanikenne et al. 2008; Wong and Cobbett 2009), using energy provided by ATP hydrolysis to transport Cd against its electrochemical gradient. The IRT1 gene, which is transcriptionally responsive to Fe deficiency, can also transport Cd to plant shoot tissues (Cohen et al. 1998), and genes from the NRAMP family are thought to play a role in Cd accumulation in Arabidopsis seedlings (Thomine et al. 2000). Certain members of the Cation Diffusion Facilitator (CDF) family function in Cd efflux or vacuolar uptake (Williams et al. 2000), while OsHMA3 has been reported to regulate xylem Cd transport by regulating vacuolar sequestration of Cd in root cells (Miyadate et al. 2011). Over-expression of HMA3 has been found to improve Cd tolerance and accumulation in Wassilewskija ecotype (Morel et al. 2009) and the gene is further thought to be responsible for differences in Cd accumulation in grains of two rice varieties (Ueno et al. 2010). Owing to its role of xylem Cd loading and transportation, HMA3 has been reported to play a key role of Cd hyperaccumulation in Arabidopsis and *N. caerulescens*. However, it is not yet well known how these metal transporters are regulated at the transcriptional level. Given the involvement of HMA3 in Cd hyperaccumulation, genetic research focusing on HMA3 may lead to new approaches to enhancing Cd accumulation and tolerance of non-hyperaccumulating plants.

Hyperaccumulators are plants that accumulate large metal concentrations in above-ground tissues (Arshad et al. 2008; Shahid et al. 2012a), have a phytoaccumulation factor index (ratio of Cd level in plant to soil) $>1$ and a translocation factor index (ratio of Cd level in plant shoot to roots) $>1$, have a high metal tolerance, and are fast growing with good biomass production. Hyperaccumulators can accumulate high levels of Cd in aerial tissues partly as a result of enhanced expression of metal transporter genes (Verbruggen et al. 2009; Santos-Echeandía et al. 2010). During the last three decades, the identification of metal hyperaccumulation traits for phytoremediation has been the focus of intensive research (Niazi et al. 2011c, 2012; Llugany et al. 2012; Sabir et al. 2015). To date, more than 500 hyperaccumulating species (belonging to 45 families) have been reported worldwide, yet the number that tolerate and accumulate Cd specifically remains largely unknown (Zhou and Song 2004). So far, identified Cd hyperaccumulators include *T. caerulescens* (Baker et al. 1994), *Arabidopsis halleri* (Dahmani-Muller et al. 2001), *Solanum nigrum* (Xiao et al. 2010), *S. alfredii* (Sun et al. 2007), *Arabis paniculata* (Qiu et al. 2008), *S. nigrum* (Sun et al. 2008), *Lonicera japonica* (Liu et al. 2009b), *Bidens pilosa* (Sun et al. 2009), *Arthrocnemum macrostachyum* (Redondo-Gómez et al. 2010) and *Viola baoshanensis* (Wu et al. 2010). These plants can solubilize soil Cd through root exudates (Redondo-Gómez et al. 2010), thereby enhancing Cd uptake and translocation.

Increased accumulation of Cd especially in aerial parts of hyperaccumulator plants is highly useful for the remediation of Cd-polluted sites. Hyperaccumulators generally have well-established detoxification mechanisms and can therefore accumulate high levels of Cd without showing any physio-morphological toxicity symptoms. Some hyperaccumulators are native to areas with high soil metal concentration, and thus grow well in soils with high Cd levels. On the other hand, plants that do not efficiently cope with Cd and other pollutants are termed as 'sensitive'. These plants do not possess well-established detoxification mechanisms and are often used in risk assessment studies. In fact, metal exposure to these sensitive plants (such as *V. faba*), even at low levels, can induce several toxic effects resulting in reduced plant growth (Shahid et al. 2014d).

## 4.5 *Cadmium Accumulation in Edible Plant Parts*

Cadmium translocation to plant parts that are animal feed and/or food products (fruit) may pose serious health concerns. Therefore, there is increasing concern regarding food safety due to environmental pollution by Cd, especially since the *Itai–Itai* disease was reported in the late 1950s (Hagin and Kono 1955; Tang et al. 2016). For populations not subjected to a Cd-contaminated environment, consumption of Cd contaminated food (vegetables, cereals, pulses and legumes) is the main source of Cd exposure in animals and humans (Järup and Akesson 2009; Mombo et al. 2016). In the United States (US), the consumption of contaminated food is the largest source of Cd intake in children and non-smoking adults (Tavarez et al. 2015). It is estimated that in the US, the average Cd consumption via contaminated food (primarily cereal grain and vegetables) is between 8 and 30 μg of Cd per day per person (Järup and Akesson 2009; Schwartz and Reis 2000; Clemens et al. 2013). Human exposure to excess Cd via contaminated food causes severe harm to organs such as kidney, liver and lungs, which elevates the risk of developing cancer (Järup and Akesson 2009; Shahid et al. 2014c; Baldantoni et al. 2016). For these reasons, Cd uptake and accumulation in edible plants and its possible effects on human health have received considerable attention (Garate et al. 1993; Akoumianakis et al. 2008; Mombo et al. 2016).

Since Cd is toxic to humans at lower concentrations than plants, plants that appear healthy are not necessarily safe for human consumption. To decrease the potential risks associated with excessive Cd intake via consumption of contaminated food, maximum allowable values of Cd in edible plant parts have been established by the Codex Alimentarius Commission of Food and Agriculture Organization of World Health Organization (FAO/WHO, CODEX 2006) (Table 2). Threshold values of Cd in edible plant parts vary with food type (cereal, legumes or

**Table 2** Threshold values of Cd in edible plant parts established by the Codex Alimentarius Commission of FAO/WHO (CODEX 2006)

| Food | Threshold values (mg/kg) | Remarks |
|---|---|---|
| Cereals, Pulses and Legumes | 0.1[a] | Excluding Bran and Germ, Wheat Grain, Rice, Soybeans and Peanuts |
| Wheat Grain and Rice | 0.2[b] | Including Bran and Germ |
| Soybeans and Peanuts | 0.2[b] | |
| Vegetable, including potatoes (edible part) | 0.5[b] | Excluding Leafy Vegetables, Fresh Herbs, Stem and Root Vegetables, Fungi, Tomatoes and Peeled Potatoes |
| Peeled Potatoes, Stem and Root Vegetables | 0.1[b] | Excluding Celeriac |
| Leafy Vegetables, Fresh Herbs, Fungi and Celeriac | 0.2[b] | |

[a]Indicates guideline level
[b]Indicates maximum level

vegetables) as well as the edible/consumed plant part (grain, root, stem or leaf) (Table 2). The tolerable limit of Cd intake by humans is < 70 μg day$^{-1}$ (FAO/WHO 1978), and the provisional tolerable weekly intake is 7 μg/kg body weight. Several recent studies have reported higher accumulation of Cd in edible plant parts than the maximum allowable values set by the Codex Alimentarius Commission—thus indicating a potential human health threat (Table 2). Among plants consumed by humans, mushrooms, rice, lettuce and other leafy vegetables are reported to be the main source of Cd exposure (Table 3). Recently, Schlecht and Ina (2015) found that that 54 % of mushrooms collected from different locations near high traffic areas in Berlin-Germany had Cd levels that exceeded the threshold values. However, it must be kept in mind that Cd uptake, translocation and compartmentation (in edible or other plant parts) in crops vary greatly among cultivars within the same species (Simmons et al. 2005; Marmiroli et al. 2014). For example Yu et al. (2006) reported high variation of Cd accumulation in rice grains (0.30–2.19 mg kg$^{-1}$) among 43 cultivars when grown on Cd contaminated soil. Shi et al. (2015) also showed variation in Cd concentration (0.14–0.22 mg kg$^{-1}$) in rice grains among 12 cultivars. Therefore, in addition to crop type, cultivar and species type must be considered in risk assessment studies and in setting food consumption guidelines.

## 4.6 *Foliar Uptake of Cd*

Nowadays, metal recycling and production processes release a large proportion of fine and ultra-fine metallic particulate matter (PM) containing heavy metals into the atmosphere due to the use of thinner and more effective filters (Schreck et al. 2014). Since these fine particles containing Cd and other heavy metals are readily transported over long distances, biomonitoring studies are currently being used to assess metal pollution via atmospheric deposition. Health risks associated with atmospheric Cd pollution arise mainly from inhalation of particles and consumption of polluted vegetables (Morman and Plumlee 2013). Although root-metal uptake is considered the main uptake pathway for plants, foliar uptake has also been identified as a potentially significant route of Cd exposure (Shahid et al. 2013b; Schreck et al. 2014; Xiong et al. 2014a, b; Uzu et al. 2010). Recently, Schreck et al. (2013) suggested that root and shoot metal transfer pathways are independent, with an additive effect in terms of toxicity. The idea of foliar metal uptake grew in the 1950s subsequent to the comment by Michigan State University that "a leaf is a very efficient organ of absorption". However, unlike Cd uptake by roots, which has been widely studied (Vaculík et al. 2012), comparatively few studies have investigated the uptake of atmospherically deposited metals by leaves (Uzu et al. 2010).

While foliar uptake of Cd is generally negligible, under certain situations (especially near mines), foliar uptake should be considered (Schreck et al. 2012a). As accumulation is roughly a function of the deposited amount of pollutants, the exposure time and compound effects of climatic factors are of prime importance. In experimental work using labelled Cd in the soil to distinguish between soil and

**Table 3** Cadmium concentrations in edible plant parts of different vegetables and crops, reported in studies published between 2011 and 2015

| Crop | Cd exposure level (mg/kg) | Cd in edible plant part (mg/kg) | Medium | References |
|---|---|---|---|---|
| *Lettuce* | 0.8 | 0.73 | Soil | Mombo et al. (2016) |
| *Celery* | 0.8 | 0.4 | Soil | Mombo et al. (2016) |
| *Celeriac* | 0.8 | 0.7 | Soil | Mombo et al. (2016) |
| *Chard* | 0.6 | 0.4 | Soil | Mombo et al. (2016) |
| *Lactuca sativa* | 42 | 9 | Soil | Baldantoni et al. (2016) |
| *Cichorium endivia* | 42 | 4.5 | Soil | Baldantoni et al. (2016) |
| *Lettuce* | 1.87 | 3.62 | Soil | Tang et al. (2016) |
| *Cacao beans* | 2.6 | 3 | Soil | Chavez et al. (2015) |
| *Solanum lycopersicum* | 100 μM | 13 | Soil | Hédiji et al. (2015) |
| *Mushrooms* | 5.23 | 2.88 | Soil | Liu et al. (2015) |
| *Alchemilla xanthochlora* | 16.3 | 8 | Soil | Quezada-Hinojosa et al. (2015) |
| *Cynosurus cristatus* | 16.3 | 9 | Soil | Quezada-Hinojosa et al. (2015) |
| *Hypericum maculatum* | 16.3 | 3 | Soil | Quezada-Hinojosa et al. (2015) |
| *Wild edible mushrooms* | NM | 10 | Soil | Schlecht and Saumel (2015) |
| *Wheat* | 13.7 | 0.22 | Soil | Shi et al. (2015) |
| *Brassica napus* | 0.3 | 6 | Soil | Wu et al. (2015) |
| *Brown rice* | 0.1 mmol/L | 3.4 | Hydroponics | Xue et al. (2014) |
| *Lettuce* | NM | 1.8 | Foliar | Xiong et al. (2014) |
| *Radish* | NM | 1.6 | Foliar | Xiong et al. (2014) |
| *Parsley* | NM | 0.9 | Foliar | Xiong et al. (2014) |
| *Brassica campestris L.* | 1.59–5.18 | 1.08 | Soil | Mahmood and Malik (2014) |
| *Brassica rapa L.* | 1.59–5.18 | 0.36 | Soil | Mahmood and Malik (2014) |
| *Spinacia oleracea L.* | 1.59–5.18 | 1.9 | Soil | Mahmood and Malik (2014) |
| *Green vegetables* | 0.52–1.61 | 1 | Soil | Chen et al. (2014a, b) |
| *Garlic chives* | 0.52–1.61 | 0.5 | Soil | Chen et al. (2014a, b) |
| *Spinach* | 0.52–1.61 | 1.1 | Soil | Chen et al. (2014a, b) |
| *Chinese cabbage* | 0.52–1.61 | 0.55 | Soil | Chen et al. (2014a, b) |
| *Garden pea* | 0.52–1.61 | 0.55 | Soil | Chen et al. (2014a, b) |
| *Rice grains* | 6 $mg/cm^3$ | 7 | Soil | Pereira et al. (2011) |
| *Lettuce* | 6 $mg/cm^3$ | 130 | Soil | Pereira et al. (2011) |

*NM* not mentioned

airborne Cd in several crops, the the impact of airborne Cd was not negligible, even in rural areas (Hovmand et al. 1983). Harrison and Chirgawi (1989) used growth cabinets with filtered air and demonstrated that the atmospheric contribution to the contamination of vegetables was as large as 23 % for Cd.

While more research is needed to further elucidate the mechanisms involved in metal foliar transfer (Zangi and Filella 2012), the two major mechanisms involved in foliar Cd uptake are thought to include (1) deposition and internalization through the lipophilic cuticle and (2) penetration through stomata openings (Schreck et al. 2012a, Barber et al. 2004). According to Chamel et al. (1991), penetration through the cuticle would consist of four steps: sorption in the cuticle, diffusion through the cuticle, desorption in the apoplast and absorption by the subjacent cells. After this penetration phase, active transport within cells (symplastic pathway), transport in the vascular system (phloem or xylem) and transport between the phloem and xylem intervene. Metal uptake via root or shoot transfer may have different effects on metal localisation in different plant tissues, metal speciation (Sarret et al. 2013), and phytotoxicity. Moreover, plant response (detoxification and tolerance mechanisms) may also vary with the type of Cd exposure (root or shoot). Additional studies are required to compare the two pathways of Cd accumulation (i.e., root or shoot), and subsequent speciation, localization, toxicity and detoxification in plant tissues—especially near industrial and mining areas.

The processes of foliar metal uptake and cuticular penetration strictly depend on a number of factors, such as the physicochemical properties of the compound and cuticle, the surface area and morphology of the plant leaf, surface texture (roughness and pubescence), plant habitus (evergreen or deciduous), the physical and chemical forms of the deposited element, duration of exposure, gas exchange and environmental conditions (Uzu et al. 2010; McLaughlin et al. 2011; Calderón-Preciado et al. 2013). The adsorption of Cd on plant foliage depends on its concentration in air as particulate matter (PM). Climatic conditions also influence the efficacy of foliar Cd transfer through direct effects on the physico-chemical properties of the leaf surface, and by affecting biological processes in the plant, both over the short- and long-term. The immediate conditions of light, temperature and humidity at the time of foliar Cd application affect plant metabolic status and hence may directly influence absorption processes across the leaf surface and from within the internal leaf spaces. Schreck et al. (2012a) found metal accumulation in lettuce shoots to be linked to precipitation (correlation coefficient: 0.83), with high metal accumulation occurring during periods of greatest rainfall. This was possibly due to enlargement of the cuticle, pores, and stomata openings (Schreck et al. 2012a) which would have facilitated leaf penetration by the metal. However, climatic processes such as rainfall and wind can also remove Cd from shoot surfaces (Rodrigo et al. 1999; Staelens et al. 2008). In the case of vegetables, washing of samples prior to analysis resulted in a 25 % decrease in foliar metal concentration (Schreck et al. 2012a).

# 5 Toxic Effects of Cd on Plants

The processes related to Cd toxicity in plants are complex (Bandyopadhyay and Mukherjee 2011). Cadmium accumulation in plants, even at low levels, adversely affects plant growth and morphology (Hasan et al. 2009; Saidi et al. 2014; Shahid et al. 2015b). Above the toxic threshold, Cd can negatively affect morphological, physiological, and biochemical functions (Benavides et al. 2005; Deng et al. 2014; Sergeant et al. 2014). Cadmium may cause plant death when its level increases above 5–10 mg Cd $g^{-1}$ leaf dry weight (Gill et al. 2013). Data summarized in Table 4 (published after 2010) shows toxic effects of Cd in different plant species under varous applied levels of Cd. The key symptoms of Cd-induced toxicity in plants are chlorosis, inhibition of photosynthesis, stunted growth, inactivation of enzymes in $CO_2$ fixation, root elongation, altered chloroplast ultrastructure, impared seedling development, cell division and transpiration, leaf epinasty, lipid peroxidation, and disturbance of sulfur and nitrogen metabolism (Das et al. 1998; Benavides et al. 2005; Gill et al. 2013; Ran et al. 2014).

Cadmium has also been reported to reduce the uptake and transport of essential nutrients (phosphorus, calcium, magnesium and potassium) (Das et al. 1998), and to interfere with nitrate uptake and translocation to aerial parts (Hernandez et al. 1996) via the reduction of nitrate reductase activity (Mathys 1975). Cadmium toxicity can influence plasma membrane permeability by inducing lipid peroxidation (Fodor et al. 1995). At the cellular level, Cd causes increased generation of reactive oxygen species (ROS), interrupts cell redox status, deteriorates biological macromolecules (lipids, proteins and nucleic acids), dismantles membranes, and cleaves DNA-strands (Shahid et al. 2014f). The Cd-induced changes in biomembranes alter their permeability to protons, nutrient and water, and also affect the activity of membrane-bound enzymes (Belkadhi et al. 2015).

## 5.1 *Effects of Cd on Seed Germination and Plant Growth*

Although the reason is not yet well understood, one of the major consequences of the cumulative impacts of Cd exposure is a decrease in seed germination. This has been reported in Cd-exposed *Lycopersicon esculentum*, *Cucumis sativus*, *Lactuca sativa*, *D. carota*, *O. sativa*, *Z. mays*, *Hordeum vulgare*, *Pinus halepensis*, *Elsholtzia argyi*, *Spartiana alterniflora* and *T. aestivum* (Wang et al. 2012; Irfan et al. 2014; Ehsan et al. 2014).

Since Cd has a tendency to preferentially accumulate in the roots (Gill et al. 2012) it is not surprising that root growth and anatomy (Lux et al. 2010) are two characteristics adversely affected by Cd exposure. At the biochemical level, Cd has been shown to induce root cell wall lignification (Vaculík et al. 2012), premature lignification and xylogenesis (Ďurčeková et al. 2007), root browning, and reduced root length (Cherif et al. 2011; Wang et al. 2012) (Table 4). However,

**Table 4** Toxic effects of Cd on plant growth. Data is collected from manuscripts published between 2010 and 2015

| Cd exposure level | Duration (days) | Cd toxicity | Plant species | Growth medium | References |
|---|---|---|---|---|---|
| 500 μM | 7 | Decreased plant biomass and photosynthetic pigments | *Gossypium hirsutum* | Hydroponics | Daud et al. (2015) |
| 100, 500 μM | 8, 13, 25 | Decreased plant growth, chlorophyll, nitrates and iron content | *Nicotiana tabacum* | Hydroponics | Iannone et al. (2015) |
| 200 mg/kg | 30 | Reduced Rubisco activity, chlorophyll contents, gas exchange parameters, leaf nitrogen content and plant growth | *Triticum aestivum* | Soil | Khan et al. (2015) |
| 1000 μM | 45 | caused electrolyte leakage, decreased plant biomass, relative water content, free proline, total soluble sugars and starch concentrations | *Triticum aestivum* | Hydroponics | Rady and Hemida (2015) |
| 50 μM | 1 h | Decreased quantum yields of PSII and protein contents | *Populus nigra* | Hydroponics | Lomaglio et al. (2015) |
| 50 μM | 14 | Plant height, root length leaf area, and number of leaves | *Brassica napus* | Hydroponics | Ehsan et al. (2014) |
| 1 mM | 7, 14 | Root and shoot length, leaf area per plant, and plant dry mass, number of pods per pot, pod yield per pot, and pod protein | *Phaseolus vulgaris* | Hydroponics | Howladar et al. (2014) |
| 20 μM | 4 | Root and leaf fresh weight, chlorophyll and carotenoid content | *Helianthus annuus* | Hydroponics | Saidi et al. (2014) |
| 120, 240, 480 μM | 30, 60 | Plant length, fresh mass, dry root and shoot mass, and leaf area | *Brassica juncea* | Soil | Irfan et al. (2014) |
| 50 mM | 30 | Rubisco activity and net photosynthesis | *Brassica juncea* | Hydroponics | Asgher et al. (2014) |

(continued)

**Table 4** (continued)

| Cd exposure level | Duration (days) | Cd toxicity | Plant species | Growth medium | References |
|---|---|---|---|---|---|
| 5, 10, 20, 40 μM | 7 | Root mass, leaf mass, total mass, and leaf width. | *Ceratopteris pteridoides* | Hydroponics | Deng et al. (2014) |
| 20 μM | 56 | Plant growth. Necrosis of youngest leaves was also observed. | *Populus tremula* | Hydroponics | Sergeant et al. (2014) |
| 100, 500 mM | 15 | Fresh and dry biomass, root diameter, root surface area, number of root tips, and root volume. | *Brassica napus* | Hydroponics | Ali et al. (2013) |
| 500, 1000 μM | 56 | Shoot length, leaf area per plant, and fresh and dry plant biomass. | *Triticum aestivum* | Hydroponics | Agami and Mohamed (2013) |
| 1, 5 μM | 35 | Plant height, root length, number of leaves per plant, and leaf area. | *Gossypium hirsutum* | Hydroponics | Farooq et al. (2013) |
| 30 μmol/L | 7 | Root and shoot length, and root and shoot biomass. | *Hibiscus cannabinus* | Hydroponics | Feng-tao et al. (2013) |
| 15, 29, 43, 58 μM | 190 | Shoot and root length, fresh and dry mass, leaf area, chlorophyll content, and photosynthesis-related traits. | *Solanum lycopersicum* | Hydroponics | Hayat et al. (2012) |
| 50 or 200 mM | 7 | Shoot length, root and shoot fresh weight, and chlorophyll a and b content. Chlorosis was also observed. | *Brassica juncea* | Hydroponics | Mohamed et al. (2012) |
| 50–100 μM | 15 | Chlorophyll and carotenoid levels. | *Arabidopsis thaliana* | Hydroponics | Martínez-Peñalver et al. (2012) |
| 36 μM | 1, 2, 3, 4, 7, 14 | Increased lateral root formation, cell-to-cell separation events leading to schizogenous oil duct formation. | *Daucus carota* | Hydroponics | Toppi et al. (2012) |

(continued)

**Table 4** (continued)

| Cd exposure level | Duration (days) | Cd toxicity | Plant species | Growth medium | References |
|---|---|---|---|---|---|
| 120, 241, 483 μM | 30 | Plant growth, net photosynethetic rate, stomatal conductance, intercellular $CO_2$, chlorophyll content. | *Lepidium sativum* | Hydroponics | Gill et al. (2012) |
| 10 μM | 27 | chlorophyll content, xylem-vessel number, and diameter of mature leaves. Plant death, water and gaseous losses in mature leaves were also observed. | *Myriophyllum alterniflorum* | Hydroponics | Delmail et al. (2011) |
| 10 μmol/L | 7 | Phytosynthetic pigmentation. | *Solanum lycopersicum* | Hydroponics | Cherif et al. (2011) |
| 483, 966 μM | 10 | Root and shoot fresh weight. | *Brassica juncea* | Hydroponics | Ahmad et al. (2011) |
| 15, 29, 43, 58 μM | 60 | Chlorophyll content, transpiration rate, stomatal conductance, and internal $CO_2$ concentration. | *Solanum lycopersicum* | Hydroponics | Hasan et al. (2011) |
| 10, 25, 50 μM | 14 | Root and shoot fresh and dry weight, biomass, growth rate, and chlorphyll content. | *Tagetes patula* | Hydroponics | Liu et al. (2011) |
| 0–97 μM | 7 | Photosynethic pigmentation, chlorophyll a and b, total chlorophyll ratios. | *Groenlandia densa* | Hydroponics | Yılmaz and Parlak (2011) |
| 200–500 μM | 14 | Fresh and dry biomass. | *Juncus effusus* | Hydroponics | Najeeb et al. (2011) |
| 0.5 mM | 1, 3, 7 | Plant growth | *Solanum lycopersicum* | Hydroponics | Monteiro et al. (2011) |
| 400 μM | 14 | Root elongation and plasma membrane integrity. | *Sedum alfredii* | Hydroponics | Tian et al. (2011) |
| 0–30 μM | 45 | Root and shoot growth. | *Solanum nigrum* | Hydroponics | Fidalgo et al. (2011) |
| 5 μM | 1, 5, 10, 15, 25 | Plant height, root length, and biomass. Leaf necrosis and root browning were also observed. | *Hordeum vulgare* | Hydroponics | Chen et al. (2010) |

(continued)

**Table 4** (continued)

| Cd exposure level | Duration (days) | Cd toxicity | Plant species | Growth medium | References |
|---|---|---|---|---|---|
| 50, 100, 200 μM | 28 | Plant growth, plant height, number of tillers, water content, and chlorophyll content. | *Achnatherum inebrians* | Hydroponics | Zhang et al. (2010) |

most studies carried out regarding risk assessment and Cd toxicity to plants after 2010 were carried out using a hydroponic (as opposed to soil) medium (Table 4).

At lower exposure levels (e.g., 0.5 μM), Cd has been reported to enhance the production of root hairs proximal to the root apex in various plants (Kopittke et al. 2010). However, the hindrance of root hair production has also been reported at high exposure levels (e.g., 0.5 mM) (Gratão et al. 2009). Increased root diameter can also result from Cd exposure (Maksimović et al. 2007; Lux et al. 2011), possibly attributed to an increase in parenchyma cell size, which enhances the resistance to radial flows of water and solutes (Maksimović et al. 2007).

In addition to effects on roots, Cd exposure is also known to negatively affect plant growth and biomass (Rady 2011) (Table 4). Based on the results of light microscopy work with *D. carota*, Corguinha et al. (2012) proposed that Cd affects plant growth in two separate phases, with the first phase impacting the functional and cytological events (1–4 days of Cd exposure), and the second phase impacting cell to cell separation, cell hypertrophy, and thiol-peptide content (4–14 days of Cd exposure). Cadmium is also known to inhibit photosynthesis by a variety of means including: reducing the transcription of genes which are related photosynthesis such as rbcL, psaB and psbA (Qian et al. 2010); inactivating Rubisco and phosphoenol pyruvate carboxylase (Bashir et al. 2013); inducing lipid peroxidation (Iannone et al. 2010), interfering with protochlorophyllide production (Prasad 2004), reducing the activity of enzymes involved in $CO_2$ fixation and inhibiting chlorophyll biosynthesis (Stobart et al. 1985); enhancing proteolysis (Pena et al. 2007); interfering with leaf transpiration through changes in stomatal conductance and stomatal density (Souza et al. 2011; Baryla et al. 2001) and; disturbing nutrient metabolism and plant antioxidant machinery (Gill and Tuteja 2010). For *B. napus*, Cd application has also been found to reduce the concentration of leaf pigments, including lutein, neoxanthin, β-carotene and violaxanthin (Baryla et al. 2001).

Generally, photosystem II (PS II) is more affected by Cd toxicity than photosystem I (PS I) (Baker, 1991). However, the impact of Cd toxicity on PS systems I and II is not still clear. In a time course experiment, Herbette et al. (2006) revealed that Cd downregulated many genes involved in glucosinolate biosynthesis and photosynthesis within hours of Cd exposure, while genes involved in phenylpropanoid metabolism, cell wall metabolism and sulfur metabolism were rapidly induced. Ultimately, high levels of Cd exposure, as well as long term exposure of elevated Cd, may cause plant death (Chen et al. 2010; Mohamed et al. 2012).

## 5.2 *Cadmium Genotoxicity*

Cadmium has the capacity to affect plant genetic material (Liu et al. 2012; Andresen and Küpper 2013; Xie et al. 2014) in a variety of ways, including the reduction of nuclear DNA content (Monteiro et al. 2012), and DNA damage that includes modified bases, single- and double-strand breaks, DNA-protein cross-links, 8-hydroxyguanine, DNA demethylation and base oxidation (Liu et al. 2005), and sister chromatid exchange (Unyayar et al. 2006). Cadmium-induced DNA damage may occur indirectly by activating and deactivating levels of radical scavengers, intracellular enzymatic processes, and the defense mechanism of target cell. It is believed that Cd damage to DNA is associated with the inhibition of 8-oxoguanine-DNA glycosylase activity (Kovalchuk et al. 2005). Cadmium may be transported across cellular/nuclear membranes and disrupt the DNA repair and replication mechanisms by binding directly or indirectly to DNA and/or protein (Liu et al. 2012). Cadmium exposure may also cause the breakdown of 1300 bp, a ladder-like degradation of DNA (Andosch et al. 2012). Some authors, however, believe that Cd alone could not induce DNA damage, but that it insteaed acts in combination with other genotoxic substances (Bandyopadhyay and Mukherjee 2011).

Cadmium can cause DNA damage/oxidation indirectly via ROS production. The hydroxyl radical ($HO^{\bullet}$) is highly electrophilic and can oxidize DNA by reacting with pyrimidine aromatic rings. The $HO^{\bullet}$ radical preferentially attacks position 5 of cytosine and thymine and can also attack the hydrogen atom of the thymine methyl group, resulting in the formation of an allylic radical (which can in turn form peroxyle radicals by reacting with oxygen) (Nzengue et al. 2015). The peroxyl radicals are dismuted to hydroperoxides, which can induce the protonation reaction (Tremblay et al. 1999), mainly in the presence of metals. The purine bases of DNA are more perceptive to Cd-induced ROS, particularly those with lower oxidation potential. Generally, ROS induced formation of 8-OHdG starts by transferring an electron to guanine, resulting in the formation of a radical cation. This radical cation loses a proton to generate a neutral radical, which acts as an oxazolone precursor. Alternatively, the radical formed from guanine may produce a neutral radical intermediate by hydration, which is reduced or oxidized respectively to generate 2,6-diamino-4-hydroxy-5-formamidopyrimidine and 8-OHdG (Nzengue et al. 2015).

In the absence of DNA repair, the formation of 8-OHdG may result in the GC to TA transversion, which is involved in the development of cancer (Feng et al. 2003). Formation of 8-OHdG is often used as biomarker of DNA damage (Nzengue et al. 2015). Cd-induced ROS can also cause DNA damage indirectly by reacting with proteins and generating 8-OHdG (Luxford et al. 2002). For example, peroxyle radicals of protein can cause oxidative lesions like 8-OHdG, whereas alkoxyl radicals of protein can generate DNA adducts (Luxford et al. 2002). DNA damage can also occur as a result of Cd-induced lipid peroxidation, since by-products of lipid peroxidations such as 4-hydroxy-2-nonenal and malonedialdehyde are capable of reacting with DNA bases (Stone et al. 1990). Malonedialdehyde can react with nucleophilic atoms of cytosine, adenine and guanine by its two electrophilic carbonyl bonds (Blair 2001), which lead to the formation of a number of adducts, such as pyrimidopurinones.

Cadmium-induced DNA damage has been reported for different plant species studied using various genotoxicity techniques. Comet assays were used by Bandyopadhyay and Mukherjee 2011 for a study involving *Nicotiana tabacum* and *Allium cepa*. Liu et al. (2005 and 2009), in contrast, used random amplified polymorphic DNA (RAPD) profiles and found that Cd exposure caused loss of normal bands, appearance of new bands, and variation in band intensity. Zhang et al. (2012) used microarray analysis and observed 28 up-regulated genes and 19 down-regulated genes in Cd-exposed *O. sativa* roots. Monteiro et al. (2012) used full peak coefficient of variation (FPCV) and reported increased FPCV values by Cd in *Thlaspi arvense* and *T. caerulescens*.

Another reported symptom of Cd toxicity is the disruption of cell mitosis via micronucleus formation, chromosomal fragmentation and aberrations, loss of acentric fragments during meiosis and restrictions to mismatch repair causing mutations (Jin et al. 2003). Micronuclei emanate as a result of mitotic anomalies or chromosomal breaks that need a passage through mitosis to be recognizable (Al-Sabti and Metcalfe 1995). Cadmium-induced micronuclei are most often associated with processes involving ROS-induced redox shuttling (Circu and Aw 2010). In a 5-day experiment, Cd exposure increased micronucleus frequency in *V. faba* root tips (Foltête et al. 2012).

Cadmium may also interfere with RNA metabolism, trigger programmed cell death, cause chromatin fragmentation (Marcano et al. 2002; Behboodi and Samadi 2004), and impact genetic functioning by affecting protein structure. Protein carbonylation (or oxidative protein modification) is commonly observed after Cd exposure (Kranner and Colville 2011). In *C. sativus* seedlings, Cd exposure resulted in a two-fold increase in protein carbonylation (Gonçalves et al. 2007), which is known to inactivate proteins and cross-linking due to the conversion of side-chains of Lys, Arg, Pro and Thr to aldehyde or keto groups (Rinalducci et al. 2008).

## 5.3 Cadmium-Induced Oxidative Stress and Lipid Peroxidation

Plant organelles such as chloroplasts, peroxisomes and mitochondria are the main sites of ROS generation (~90 %) in plant cells (Lushchak 2011; Del Río 2011). Reactive oxygen species are generally short lived, unstable and chemically highly reactive molecules due to the presence of unpaired valence shell electrons (Andresen and Küpper 2013; Shahid et al. 2014f). ROS include: singlet oxygen ($^{1\!/\!2}O_2$), hydroxyl ($HO^{\bullet}$), superoxide anion ($O_2^{\bullet\,-}$), hydrogen peroxide ($H_2O_2$), peroxyl ($RO_2^{\bullet}$), alkoxyl ($RO^{\bullet}$) radicals, and organic hydroperoxide (ROOH).

Cadmium-induced ROS production is common, and one of the earliest biochemical changes exhibited by plant subjected to Cd stress (Kung et al. 2014; Cuypers et al. 2011; Andresen and Küpper 2013; Farooq et al. 2013; Pourrut et al. 2008; Shahid et al. 2014f, 2015a). Recently, Shahid et al. (2012d) reported in a time course experiment that metal-induced ROS production occurred immediately after metal concentrations reached significant levels in roots (after 1 h of exposure). Moreover, metal-induced ROS production may occur in phases (e.g., two bursts of overproduction of ROS, occurring at 1 and 12 h). Data published after 2000 regarding Cd-induced ROS production (mainly in the form of $H_2O_2$ and $O_2^{\bullet -}$) for different plant species and various exposure durations and levels of Cd are summarized in Table 5.

ROS production is known to be an unavoidable consequence of aerobic metabolism (Shahid et al. 2014d, e). Although Cd is not a redox active metal and unable to produce ROS directly via Fenton and Haber-Weiss reactions, indirect ROS production has been reported in Cd-exposed plants (Gallego et al. 1996; Hasan et al. 2009; Iannone et al. 2010). Cadmium causes three waves of ROS, i.e., hydrogen peroxide ($H_2O_2$) by NADPH oxidase, superoxide anions ($O_2^{\cdot -}$) in mitochondria and fatty acid hydroperoxide (Garnier et al. 2006). Direct interaction between Cd and enzyme binding sites resulting from the affinity of Cd for thioyl-, histidyl- and carboxyl-groups is proposed to be responsible for Cd-induced ROS production. These interactions cause enzyme inactivation which in turn shatters ROS balance and results in oxidative stress (Cuypers et al. 2011). Cadmium also causes increased ROS production via the binding and consumption of GSH and its derivatives, which are reported to restrict ROS production by binding reactive transition elements (Hodoshima et al. 2007). In addition to these, plasma-membrane-bound NADPH oxidase has been shown to be involved in Cd-induced oxidative stress for *Pisum sativum* (Rodríguez-Serrano et al. 2006), *N. tabacum* (Olmos et al. 2003; Garnier et al. 2006) and *S. nigrum* (Deng et al. 2010).

Plant NADPH oxidases are homologs of the mammalian NADPH oxidase catalytic subunit pg91phox and are called Rboh–respiratory burst oxidase homologs (Jakubowska et al. 2015). Activation of NADPH oxidase involves the translocation and association of cytoplasmic proteins with membrane bound components. The proteins involved in the activation of NADPH oxidase include two small GTP-binding proteins (Rap1A, p21rac), two plasma membrane-associated proteins (p22phox, gp91phox) and three cytosolic proteins (p47phox, p67phox, p40phox) (Pourrut et al. 2008). NADPH oxidase generates unstable superoxide radical anion ($O_2^{\bullet -}$) by catalyzing oxygen reduction using NADPH as the electron donor. Studies with diphenylene iodonium and imidazole as an inhibitor of NADPH oxidase and $NaN_3$ as an inhibitor of peroxidase revealed that NADPH oxidase and peroxidase are the main sources of Cd-induced ROS in plants (Chou et al. 2012). Similar to animals, the mitochondrial electron transfer chain is the main place of Cd-induced ROS production (Heyno et al. 2008).

The essential roles of ROS in living organisms include: programmed cell death via signal transduction, initiation of defense metabolism under stress, growth regulation, regulation of stomatal movement, fruit ripening and senescence, and

**Table 5** Cadmium-induced ROS production in different plant species. Data is collected from manuscripts published after 2000

| ROS | Plant species | Cd exposure level | Culture | Duration (days) | References |
|---|---|---|---|---|---|
| $H_2O_2$, $O_2^-$ | *Gossypium hirsutum* | 500 μM $CdCl_2$ | Hydroponics | 7 | Daud et al. (2015) |
| $H_2O_2$ | *Helianthus annuus* | 5 μM $CdCl_2$ | Hydroponics | 14 | Hawrylak-Nowak et al. (2015) |
| $H_2O_2$ | *Nicotiana tabacum* | 100, 500 μM $CdCl_2$ | Hydroponics | 8, 13, 25 | Iannone et al. (2015) |
| $H_2O_2$ | *Triticum aestivum* | 200 mg/kg $CdCl_2$ | Soil | 30 | Iqbal et al. (2015) |
| $H_2O_2$ | *Populus nigra* | 50 μM $CdSO_4$ | Hydroponics | 1 h | Lomaglio et al. (2015) |
| $H_2O_2$ | *Triticum aestivum* | 1000 μM $CdCl_2$ | Hydroponics | 45 | Rady and Hemida (2015) |
| $H_2O_2$, $O_2^-$ | *Brassica campestris* | 50 μM $CdCl_2$ | Hydroponics | 24 h | Wu et al. (2014) |
| $H_2O_2$, $O_2^-$ | *Arabidopsis thaliana* | 80 mg/L $CdCl_2$ | Hydroponics | 1, 3, 6, 12, 24 h | Gu et al. (2014) |
| $H_2O_2$ | *Brassica juncea* | 50 μM $CdCl_2$ | Hydroponics | 15 | Asgher et al. (2014) |
| $H_2O_2$ | *Ganoderma lucidum* | 1000 μM $CdCl_2$ | Hydroponics | 5 | Kung et al. (2014) |
| $H_2O_2$, $O_2^-$ | *Cynodon dactylon* | 750 mg/kg $CdCl_2$ | Soil | 21 | Shi et al. (2014) |
| $H_2O_2$ | *Arabidopsis thaliana* | 5, 10 μM $CdSO_4$ | Hydroponics | 1 | Keunen et al. (2013) |
| $H_2O_2$ | *Phaseolus vulgaris* | 20 μM $CdCl_2$ | Hydroponics | 3 | Saidi et al. (2013) |
| $H_2O_2$ | *Gossypium hirsutum* | 1, 5 μM $CdCl_2$ | Hydroponics | 35 | Farooq et al. (2013) |
| $H_2O_2$ | *Arabidopsis thaliana* | 50, 100 μM $CdCl_2$ | Hydroponics | 3, 9, 24 h | Martínez-Peñalver et al. et al. (2012) |
| $H_2O_2$ | *Chlorella vulgaris* | 100 μM Cd $(NO_3)_2$ | Hydroponics | 24 h | Piotrowska-Niczyporuk et al. (2012) |
| $H_2O_2$, $O_2^-$ | *Gracilaria dura* | 0.4 mM $CdCl_2$ | Hydroponics | 4 | Kumar et al. (2012) |
| $H_2O_2$ | *Brassica juncea* | 890, 1779 μM $CdSO_4$ | Hydroponics | 45 | Ahmad et al. (2011) |
| $H_2O_2$ | *Medicago sativa* | 750 mg/kg $CdCl_2$ | Soil | 60 | Antolín et al. (2010) |
| $H_2O_2$, $O_2^-$ | *Solanum nigrum* | 200 μM $CdCl_2$ | Hydroponics | 3 | Deng et al. (2010) |
| $H_2O_2$ | *Brassica juncea* | 200 μM $CdCl_2$ | Hydroponics | 2, 6, 12, 24 h | Guan et al. (2009) |
| $NO^•$, $H_2O_2$, $O_2^-$ | *Pisum sativum* | 50 μM $CdCl_2$ | Hydroponics | 14 | Rodríguez-Serrano et al. (2009) |

(continued)

**Table 5** (continued)

| ROS | Plant species | Cd exposure level | Culture | Duration (days) | References |
|---|---|---|---|---|---|
| $OH^{\bullet}$, $H_2O_2$, $O_2^-$ | *Ceratophyllum demersum* | 10 μM $CdCl_2$ | Hydroponics | 7 | Aravind et al. (2009) |
| $H_2O_2$ | *Triticum aestivum* | 50, 250 μM $CdCl_2$ | Hydroponics | 24 h | Singh et al. (2008a) |
| $H_2O_2$, $O_2^-$ | *Arabis paniculata* | 22, 44, 89, 178 μM $CdCl_2$ | Hydroponics | 24 h | Qiu et al. (2008) |
| $NO^{\bullet}$ | *Triticum aestivum* | 1000 μM $CdCl_2$ | Hydroponics | 5 | Groppa et al. (2008) |
| $H_2O_2$, $O_2^-$ | *Solanum tuberosum* | 100 μM $CdCl_2$ | Hydroponics | 1 h | Heyno et al. (2008) |
| $H_2O_2$ | *Vigna mungo* | 25, 50, 100 mg/kg $CdCl_2$ | Soil | 30 | Singh et al. (2008b) |
| $O_2^-$ | *Mytilus galloprovincialis* | 27–52 ng/g | Hydroponics | 7 | Koutsogiannaki et al. (2006) |
| $H_2O_2$ | *Nicotiana tabacum* | 5 mM $CdCl_2$ | Hydroponics | 7 | Olmos et al. (2003) |
| $O_2^-$ | *Lupinus luteus* | 100 μM $CdCl_2$ | Hydroponics | 6 | Kopyra and Gwóźdź (2003) |
| $H_2O_2$ | *Pisum sativum* | 50 μM $CdCl_2$ | Hydroponics | 14 | Romero-Puertas et al. (2002) |
| $O_2^-$ | *Oryza sativa* | 100, 500 μM $Cd(NO_3)_2$ | Hydroponics | 5 | Shah et al. (2001) |

$O_2^{\bullet-}$; superoxide anion, $HO^{\bullet}$; hydroxyl, $H_2O_2$; hydrogen peroxide, $NO^{\bullet}$; nitric oxide

alleviation of seed dormancy (Hasan et al. 2009; Shahid et al. 2014f and references therein). Recently, it was reported that ROS ($H_2O_2$) act as signaling molecules to emanate defense mechanisms against toxic metals including Cd. Hydrogen peroxide, for example, can activate protein kinases, which in turn regulate gene transcription by repressing or activating the transcription factors (Pandey and Somssich 2009). Several studies have reported Cd-induced activation of protein kinase transcripts via ROS accumulation in plants (Lushchak 2011; Zhang et al. 2012).

An imbalance between ROS production and elimination takes place after Cd exposure, leading to dramatic physiological challenges in the plant termed as "oxidative stress". Interaction of ROS with biomolecules and the resulting toxicity vary with the nature of target molecules and ROS. Although ROS can interact with and modify all cellular components, some are more susceptible. ROS-induced degradation/oxidation/inactivation is more pronounced for enzymes containing active thiol groups such as glyceraldehyde-3-phosphate dehydrogenase or [Fe–S]-clusters such as fumarate hydratase and aconitase, 6-phosphogluconate dehydratase (Lushchak 2011; Shahid et al. 2014f). ROS-induced degradation of most cellular constituents is irreparable. The few biomolecules that can be repaired include DNA and cysteine and methionine residues in proteins (Lushchak, 2007). A range of physiological dysfunctions are thus observed due to ROS interaction with lipids, proteins and DNA.

Recent research suggests that Cd-induced ROS causes changes in cellular redox homeostasis in favor of the pro-oxidants, mainly by activating the peroxidation of membrane lipids (Körpe and Aras 2011; Andresen and Küpper 2013). Lipid peroxidation is the process whereby ROS generated by Cd or other heavy metals oxidize lipids and their esters in cell membranes (Hasan et al. 2009; Shahid et al. 2014f). Cadmium-induced lipid peroxidation has been reported in many studies involving different plants (Liptáková et al. 2013); it occurs commonly in polyunsaturated fatty acids and involves a process of initiation, progression and termination (Shahid et al. 2014f). Lipid peroxidation results in a reduction of plant development and the generation of highly cytotoxic compounds (Márquez-García et al. 2011; Pourrut et al. 2011, 2013). This process leads to oxidative damages that compromise the general redox homeostasis (Cuypers et al. 2011) and alteration to the lipidic environment which impairs its fluidity and intrinsic membrane protein activities (Quartacci et al. 2006; Shahid et al. 2014f). ROS-induces changes to biomembranes resulting from Cd toxicity affects membrane permeability to nutrients, water and protons.

Cadmium toxicity also impacts lipid biochemistry by inhibiting biosynthetic pathways which may lead to a reduction in unsaturated fatty acids (Mohamed et al. 2012). Under Cd stress, plant species including *B. juncea*, *Capsicum annum*, *Karelian birch* and *T. aestivum* balance ROS production by regulating the composition of membrane lipids (e.g, neutrallipids, phospholipids and galactolipids) (Belkadhi et al. 2015). Moreover, Cd may substitute Ca from essential sites on the endoplasmic reticulum membranes, tonoplast, apoplast and mitochondrial (Breckle and Kahle 1991). Cadmium diffusion into cytosol activates phospholipases, which results in membrane discharge of linolenic (C18:3) and linoleic (C18:2) acids serving as substrates for ROS or lipoxygenase (Chmielowska-Bąk et al. 2014), which also has been reported in many species such as *O. sativa* and *H. vulgare* (Belkadhi et al. 2015). ROS production also has a major impact on the activation of membrane-bound enzymes ($Ca^{2+}$-ATPase and $H^+$ -ATPase).

## 6 Cadmium Detoxification in Plants

Plants naturally possess a well-organized antioxidant defense system for managing overproduction of ROS and evading oxidative stress (Anjum et al. 2012; Shahid et al. 2014f). Inherent to this defence system are mechanisms to increase: Cd elimination, Cd redistribution to insensitive tissues including sequestration in vacuoles, cellular Cd binding to cell walls, and complexation of Cd by organic ligands. These mechanisms are aided by metabolic compounds such as phytochelatins (PCs), glutathione (GSH), carotenoids and tocopherols, and enzymatic antioxidant systems including superoxide dismutases (SOD and EC 1.15.1.1), catalase (CAT and EC 1.11.1.6), peroxidase (POD, EC 1.11.1.7), ascorbate peroxidase (APX, EC 1.11.1.11), guaiacol peroxidise (GPX,, EC 1.11.1.7), glutathione reductase (GR, EC 1.6.4.2), monodehydroascorbate reductase (MDHAR, EC 1.6.5.4) and dehydroascorbate reductase (DHAR, EC 1.8.5.1). These mechanisms often act simultaneously to scavenge free ROS and can have immediate (PCs) or long-term (gene encoding) effects. However, the plasticity of induction of a specific defense mechanism and its efficiency depends on plant species and maturity, metal type and exposure duration (Andresen and Küpper 2013; Shahid et al. 2014f). Together, the increased levels of the defence-related metabolic compounds and antioxidant enzymes lead to increased stress tolerance against heavy metal-induced ROS (Pourrut et al. 2011).

## *6.1 Vacuolar Compartmentalization*

Vacuolar compartmentalization is important for Cd homeostasis and detoxification as it reduces the movement of free Cd in the cytosol (Mendoza-Cózatl et al. 2011; Choppala et al. 2014). Cadmium sequestration in vacuoles may occur via PC-dependant or -independent pathways (Wojas et al. 2009; Khoudi et al. 2012), with direct Cd transport facilitated by Cd/$H^+$ antiport activity (Salt and Wagner 1993) driven by trans-membrane $H^+$ gradients. Increased Cd sequestration in vacuoles has been reported due to the expression of CAX1 variant in petunia (Wu et al. 2011). There are several molecules involved in vacuolar Cd sequestration, but analyses from various plant species have shown that PCs and glutathione (GSH) are the main metal–ligand molecules involved.

### 6.1.1 Phytochelatins

Phytochelatins (PCs) are a family of peptides with the structure (c-Glu-Cys)n-Gly, where n represents the repetition of g-Glu–Cys, which tends to vary from 2 to 11 (Akhter et al. 2012). As with other metals, PCs play an important role in Cd detoxification (Márquez-García et al. 2011; Choppala et al. 2014). Phytochelatin synthesis under metal stress is perhaps one of the most well studied mechanisms mediating metal detoxification in plants and some yeast (Mendoza-Cózatl et al. 2011). Phytochelatins are reported to neutralize harmful effects by chelating metals with their SH groups. Of all the heavy metals, $Cd^{2+}$ has been observed to invoke the strongest activation of the phytochelatin synthase (PCS) and thus PC production (Akhter et al. 2012) due to the high affinity of the thiol groups for Cd on the cysteine residues (Chekmeneva et al. 2011). Chen et al. (2007) studied Cd–PC complexes in *Brassica chinensis* and observed binding stoichiometries of 1:1 to 3:1 depending on the availability of thiol groups and $Cd^{2+}$ in the cytosol.

The process of PC formation subsequent to Cd exposure is complex (Lima et al. 2012) and includes Cd absorption, GSH synthesis, and PCS activation followed by synthesis of the PC olygomeric chains (Ogawa et al. 2011). During PC synthesis, a g-Glu-Cys unit from GSH is first added to the GSH molecule itself and then to the elongating PC chain (Rea et al. 2004; Ramos et al. 2007). Although the PCS enzyme is thought to be self-regulated (due to the synthesis of PC following Cd exposure (Loeffler et al. 1989)), its mechanism of self-regulation remains unknown. While protein phosphorylation has been shown to prompt PCS synthesis during GSH production (Wang et al. 2009), there are no known reports of protein phosphatases involved in PC regulation. Huang et al. (2012) reported that Cd tolerance and accumulation failed when GSH and PC synthesis was blocked by L-buthionine sulfoximine, and also when PC synthesis was blocked in the PCS-deficient cad1-3 mutant.

Once formed, Cd-PC complexes are transported through the tonoplast and compartmentalized in metabolically inactive sites (Najmanova et al. 2012) such as vacuoles (Fig. 3) (Lux et al. 2011), where they are promptly sequestered (Verbruggen et al. 2009) or released to apoenzymes. Transport of the Cd-PC complex to vacuoles is thought to be facilitated by ABC transporters (ATP-binding cassette transporters) (Préveral et al. 2009), which for *O. sativa* seedlings, are encoded by OsPDR5/ABCG43 (Oda et al. 2011). A half-size ABC transporter (Hmt1; Heavy metal tolerance 1) has been identified in *Schizosaccharomyces pombe* as a vacuolar Cd-PC transporter involved in Cd tolerance (Salt and Rauser 1995). It is unclear whether Hmt1 function is GSH-dependent or PC-dependent.

Two independent groups of ABC transporters have been identified as being involved in the mediation of vacuolar PC transport in Arabidopsis (MRP1/ABCC1 and MRP2/ABCC2) and *S. pombe* (Abc2) (Mendoza-Cozatl et al. 2010). These transporters (ABCC1 and ABCC2) have been found to mediate the sequestration of metal-glutathione complexes into vacuoles (Rea 2007). Sequestration of Cd by the PC complex depends on the complex stability; with Cd–PC complexes having acid–labile sulphur possessing an improved stability and a higher Cd sequestering efficiency. Complex stability has also been found to increase with chain length up until PC3, beyond which the number of thiol groups (i.e., $\geq$4) saturate the tetrahedral coordination of $Cd^{2+}$ (Chekmeneva et al. 2011).

Moreover, PCs have also been found to be involved in the long-distance transport of Cd from root to shoot (Mendoza-Cózatl et al. 2011). Using Extended X-ray Absorption Fine Structure (EXAFS), Vogel-Mikus (2010) showed that 60 % of Cd is coordinated with PCs in *Thlaspi praecox*. This is in addition to the earlier work of Belleghem et al. (2007) who identified sulfur–Cd complexes in the cytoplasm of phloem-loading cells of *A. thaliana* (consistent with the idea that thiols (PCs) mediate long-distance Cd transport). Long-distance Cd transport via xylem sap is mediated by ZIP transporters, xylem loading Heavy Metal ATPases (HMA2 and HMA4), and PC transporters ABCC1 and ABCC2 (Mendoza-Cózatl et al. 2011). These PC-Cd complexes loaded into the phloem are more likely to be sequestered in root vacuoles (Mendoza-Cózatl et al. 2011). While the mechanisms involved in long-distance Cd translocation from root to shoot are not yet completely understood, it is possible that PC-mediated movement of Cd out of the plant shoots could result in more systemic disruption of essential processes (if the plant is unable to compartmentalize Cd and render it less biologically reactive).

Although Cd-induced production of PC in plants is well reported (Vurro et al. 2011; Ogawa et al. 2011; Son et al. 2012), some authors argue that PC is not involved at all in Cd detoxification (Inouhe et al. 2000). Ebbs et al. (2002) showed that PC synthesis had no role in Cd tolerance in *T. caerulescens*. Zhang et al. (2010) also showed no significant role of PC in Cd tolerance in *S. alfredii*. In fact, Cd is reported to induce PC synthesis only at high concentrations (Lugon-Moulin et al. 2004)—consistent with the theory of Salt and Wagner (1993) who argue that it would only be at high levels of Cd exposure that PCs might play a role. Thus the exact role of PCs in Cd detoxification for plants remains unclear. As the expression and role of PCs in Cd tolerance quite likely vary with plant type (e.g., metal sensitive or tolerant), continued research is needed to futher our understanding of the role of these compounds—particularly for plant species which can be used for phytoremediation.

### 6.1.2 Glutathione

Glutathione (GSH) is a low molecular weight tri-peptide which is the major source of non-protein thiol in plant cells (Márquez-García et al. 2011; Pourrut et al. 2013). Owing to its unique nucleophilic and redox properties, GSH plays a key role in physiological functions such as transport of GSH-conjugated amino acids and intracellular redox state regulation (Deng et al. 2010). As an important antioxidant, GSH takes part in bio-reductive reactions as an ROS scavenger to relieve cells from oxidative stress (Park et al. 2012; Bashir et al. 2013), becoming oxidized to GSSG in the process (Shahid et al. 2014f). The antioxidant properties of GSH depend on the oxidation of the SH group, giving rise to GSSG (Márquez-García et al. 2011). Glutathione synthesis takes place in two steps which are ATP-dependent. These two steps are catalyzed by glutamylcysteine synthetase (ECS) and glutathione synthetase (GS) (May et al. 1998; Fässler et al. 2011). GSHI and GSHII genes are respectively involved in encoding these two enzymes (Zhu et al. 1999). Therefore, enhanced generation of GSH due to the overexpression of g-glutamylcysteine synthetase (g-ECS) leads to increased tolerance of Cd. Overexpression of GS has also been shown to result in higher GSH levels and Cd tolerance (Zhu et al. 1999; Lugon-Moulin et al. 2004). Nowadays these genes are used in genetic engineering for increased phytoaccumulation and Cd tolerance. Under Cd stress, GSH production may decrease or increase. Glutathione contents increase during the initial metal exposure phase but then returne to control levels or lower (Chen et al. 2012). The decrease in GSH content might be due to its high rate of consumption as a reductant in antioxidant systems as metal load increases with concentration and exposure duration (Mishra et al. 2006; Chen et al. 2012).

Transgenic plants are capable of producing high GSH levels which can (but do not always) enhance tolerance to Cd and other metals (Xiang et al. 2001; Choppala et al. 2014). Indeed, GSH alone is not capable of providing enough protection against trace metal toxicity, especially under conditions of high metal loading (Noctor et al. 1998). Different physiological, biochemical and genetic studies suggest that GSH acts as a substrate for PC synthesis (Cobbett, 2000; Márquez-García et al. 2011). Glutathione is also involved in Cd-GSH transport to vacuoles via ABC transporters (Fig. 3) (Lugon-Moulin et al. 2004). Cadmium produces an imbalance of the available GSH pool because of its high affinity for thiol groups. In this way, Cd disturbs the AsA–GSH cycle which has been suggested as the source of $H_2O_2$ removal in many plants (Semane et al. 2007).

## 6.2 Antioxidant Enzymes

Cadmium is well known to enhance the activities of antioxidant enzymes—enzymes which tend to be electron donors that react with ROS to form innocuous end products (e.g., water) (Shahid et al. 2014f; Abbas et al. 2015). Several studies have reported Cd-induced activation of antioxidant enzymes (Table 6), possibly via modulation of gene expression (Anjum et al. 2012) or Cd-induced restriction of enzyme inhibitors (Ali et al. 2013; Deng et al. 2014; Chen et al. 2014a, b). Cadmium may also enhance antioxidant enzymes activity indirectly by increasing the concentration of their substrates (Anjum et al. 2012). Antioxidative enzymes activated in response to Cd stress for different plant species are highlighted in Table 6.

Because of their ability to detoxify ROS, antioxidative enzymes provide an important line of defense against Cd toxicity (Iqbal et al. 2010). The following describes some of the prinicpal known antioxidant enzymes:

*Superoxide Dismutase* (*SOD*) is a key antioxidant enzyme which defends plants against ROS by removing $O_2^{\bullet-}$ by catalyzing its dismutation (i.e., one $O_2^{\bullet-}$ is reduced to $H_2O_2$ and other oxidized to $O_2$) (Noctor and Foyer 1998; Hasan et al. 2011). $H_2O_2$, which itself may serve as a signaling molecule in the activation of antioxidative enzymes (Chou et al. 2012), is subsequently detoxified by enzymes such as GPX, APX, and CAT. Cadmium-induced activation of SOD is reported in different plant species (Saidi et al. 2014; Deng et al. 2014; Ehsan et al. 2014; Irfan et al. 2014; Howladar 2014).

*Catalayse* (*CAT*) can convert $H_2O_2$ to $O_2$ and $H_2O$ (Chen et al. 2012). CAT has very high catalytic rates but generally low substrate affinities, since the reaction requires the simultaneous access of two $H_2O_2$ molecules (Willekens et al. 1997). Several studies have shown Cd-induced increases in CAT activity in plants (Saidi et al. 2013; Chen et al. 2014a, b; Shi et al. 2014; Agami and Mohamed 2013).

*Guaiacol Peroxidase* (*GPX*) requires a reductant such as guaiacol to reduce $H_2O_2$ to $H_2O$ (Gill et al. 2012). Activation of GPX under Cd stress is reported by several authors (Monteiro et al. 2011; Wang et al. 2012; Farooq et al. 2013; Ehsan et al. 2014).

**Table 6** Cadmium-induced activation of different enzymes in different plant species. Data is collected from manuscripts published between 2010 and 2015

| Enzymes | Plant species | Cd exposure level | Culture | Duration (days) | References |
|---|---|---|---|---|---|
| SOD, CAT, POD, GR | *Gossypium hirsutum* | 500 μM $CdCl_2$ | Hydroponics | 7 | Daud et al. (2015) |
| SOD, CAT, APX, GPX | *Nicotiana tabacum* | 100, 500 μM $CdCl_2$ | Hydroponics | 8, 13, 25 | Iannone et al. (2015) |
| SOD, CAT, POD, PPO | *Zea mays and Helianthus annuus* | 10, 25 mg $kg^{-1}$ | Soil | 60 | Aghababaei and Raiesi (2015) |
| PROX, GPX, GR | *Triticum aestivum* | 200 mg/kg $CdCl_2$ | Soil | 30 | Iqbal et al. (2015) |
| GR, APX, NADPH oxidase, SOD | *Medicago sativa* | 6, 30 μM $CdCl_2$ | Hydroponics | 7 | Flores-Caceres et al. (2015) |
| SOD, CAT, APX, GR | *Trifolium repens* | 100 μM $CdCl_2$ | Hydroponics | 7 | Liu et al. (2015) |
| CAT, POX, APX | *Triticum aestivum* | 1000 μM $CdCl_2$ | Hydroponics | 45 | Rady and Hemida (2015) |
| SOD, POD, CAT | *Ricinus communis* | 2, 5 mg $L^{-1}$ $CdCl_2$ | Hydroponics | | Zhang et al. (2015) |
| GPX, DHAR, LOX, AAO | *Hordeum vulgare* | 15 μM $CdCl_2$ | Hydroponics | 3, 6 h | Tamás et al. (2015) |
| GST, CAT, GPX, APX, GR | *Triticum aestivum* | 50 μM Cd $(NO_3)_2$ | Hydroponics | 7 | Kovács et al. (2014) |
| SOD, CAT, POD, APX, GR | *Helianthus annuus* | 20 μM $CdCl_2$ | Hydroponics | 4 | Saidi et al. (2014) |
| CAT, SOD, POD | *Ceratopteris pteridoides* | 0, 5, 10, 20, 40 μM | Hydroponics | 7 | Deng et al. (2014) |
| SOD, GPX, CAT, APX | *Brassicanapus* | 10, 20 μM $CdCl_2$ | Hydroponics | 56 | Ehsan et al. (2014) |
| SOD, CAT, POX | *Brassica juncea* | 25, 50,100 mg/kg $CdCl_2$ | Soil | 30, 60 | Irfan et al. (2014) |
| SOD, POD, GR, CAT | *Phaseolus vulgaris* | 90 μM, 1 M $CdCl_2$ | Soil | 35 | Howladar (2014) |
| SOD, POD, CAT | *Phanerochaete chrysosporium* | 1–500 μM Cd $(NO_3)_2$ | Hydroponics | 1 | Chen et al. (2014a, b) |
| CAT, APX, GPX | *Brassica juncea* | 5, 35 mg/kg $CdCl_2$ | Soil | 60 | Armas et al. (2014) |
| SOD, CAT, POD, GR | *Cynodon dactylon* | 750 mg/kg $CdCl_2$ | Soil | 21 | Shi et al. (2014) |

(continued)

**Table 6** (continued)

| Enzymes | Plant species | Cd exposure level | Culture | Duration (days) | References |
|---|---|---|---|---|---|
| SOD, CAT, POX | *Triticum aestivum* | 500, 1000 μM $CdCl_2$ | Hydroponics | 56 | Agami and Mohamed (2013) |
| CAT, SOD, POD | *Arabidopsis thaliana* | 150 μM $CdCl_2$ | Hydroponics | 7 | Tao et al. (2013) |
| CAT, APX | *Phaseolus vulgaris* | 20 μM $CdCl_2$ | Hydroponics | 3 | Saidi et al. (2013) |
| SOD, GPX, CAT, APX | *Gossypium hirsutum* | 1, 5 μM $CdCl_2$ | Hydroponics | 35 | Farooq et al. (2013) |
| APX, SOD, POD | *Brassica napus* | 100 and 500 μM $CdCl_2$ | Hydroponics | 15 | Ali et al. (2013) |
| SOD, GR, APX, CAT | *Arabidopsis thaliana* | 50 μM $CdCl_2$ | Hydroponics | 3, 5 | Bashir et al. (2013) |
| SOD, CAT, POD, GR | *Hibiscus cannabinus* | 20, 50, 80, 120 μM $CdCl_2$ | Hydroponics | 21 | Feng-tao et al. (2013) |
| SOD, CAT, GR, GST, APX | *Lemna gibba* | 0.05, 0.5, 5, 10, 20 $mgL^{-1}$ Cd $(NO_3)_2$ | Hydroponics | 7 | Parlak and Yilmaz (2013) |
| SOD, APX, GR | *Gracilaria dura* | 0.4 mM $CdCl_2$ | Hydroponics | 4 | Kumar et al. (2012) |
| APX, MDHAR, DHAR, GR, GST | *Helianthus annuus* | 10 mg/kg Cd | Soil | 3 | Nehnevajova et al. (2012) |
| SOD, GR, APX | *Lepidium sativum* | 25, 50, 100 mg $kg^{-1}$ | Soil | 30 | Gill et al. (2012) |
| POD, APX, CAT, SOD | *Triticum aestivum* | 0.10 mmol/l $CdCl_2$ | Hydroponics | 10 | Guo et al. (2012) |
| SOD, POX, APX, CA | *Solanum lycopersicum* | 3, 6, 9, 12 mg/kg Cd | Soil | 20 | Hayat et al. (2012) |
| CAT, POX | *Brassica juncea* | 50, 200 μM Cd $(NO_3)_2$ | Hydroponics | 7 | Mohamed et al. (2012) |
| SOD, GR, APX | *Erica andevalensis* | 0.53, 5.3, 53, 530 μM $CdCl_2$ | Hydroponics | 5 | Márquez-García et al. (2012) |
| SOD, GPOX, APX, CAT | *Vicia faba* | 6 mmol/l $CdCl_2$ | Hydroponics | 15 | Wang et al. (2012) |
| SOD, GR, APX, CAT, GST | *Groenlandia densa* | 0.05, 0.5, 5, 10, 20 $mgL^{-1}$ Cd $(NO_3)_2$ | Hydroponics | 7 | Yılmaz and Parlak (2011) |
| SOD, POD, CAT, APX, GR | *Juncus effusus* | 10, 50 and 100 μM Cd $(NO_3)_2$ | Hydroponics | 14 | Najeeb et al. (2011) |

(continued)

**Table 6** (continued)

| Enzymes | Plant species | Cd exposure level | Culture | Duration (days) | References |
|---|---|---|---|---|---|
| SOD, SOD, CAT, APX, GR, SPX | *Arabidopsis thaliana* | 5 μM Cd | Hydroponics | 3 | Vanhoudt et al. (2011) |
| SOD, GPX, CAT, APX, GR, | *Solanum lycopersicum* | 0.5 mM $CdCl_2$ | Hydroponics | 7, 20, 36 | Monteiro et al. (2011) |
| SOD, CAT, GPOX | *Sedum alfredii* | 400 μM Cd | Hydroponics | 1 | Tian et al. (2011) |
| APX, GR, SOD, APX | *Tagetes patula* | 10, 20, 50 μM $CdCl_2$ | Hydroponics | 2 | Liu et al. (2011) |
| SOD, CAT, APX, GR | *Solanum lycopersicum* | 10 μM $CdCl_2$ | Hydroponics | 7 | Cherif et al. (2011) |
| APX, MDHAR, DHAR, SOD, GST | *Typha latifolia* | 10, 50, 100, 250 μM $CdCl_2$ | Hydroponics | 3 | Lyubenova and Schröder (2011) |
| SOD, APX, CAT, GR | *Brassica juncea* | 890, 1779 μM $CdSO_4$ | Hydroponics | 45 | Ahmad et al. (2011) |
| SOD, CAT, APX | *Solanum nigrum* | 7.5, 15, 30 μM $CdSO_4$ | Hydroponics | 45 | Fidalgo et al. (2011) |
| SOD, POX, APX, CA | *Solanum lycopersicum* | 3, 6, 9,12 mg kg $^{-1}$ | Soil | 60 | Hasan et al. (2011) |
| POD, CAT | *Amaranthus hybridus* | 30, 60, 90, 120, 150, 180 mg $kg^{-1}$ $CdCl_2$ | Soil | 90 | Zhang et al. (2010) |
| SOD, GR, APX | *Atriplex halimus* | 100 μM $CdCl_2$ | Hydroponics | 112 | Lefèvre et al. (2010) |

*SOD* superoxide dismutase, *APX* ascorbate peroxidise, *GPX* guaiacol peroxidase, *CAT* catalase, *PPO* polyphenol oxidase, *GR* glutathione reductase, *AsA* ascorbic acid, *GSH* glutathione, *GST* Glutathione S-transferase, *ACOX* acyl co-A oxidase, *POD* peroxidase, *DHAR* dehydroascorbate reductase, *MDHAR* monodehydroascorbate reductase, *LOX* lipoxygenase, *AAO* ascorbic acid oxidase, *SDH* succinatedehydrogenase, *PROX* proline oxidase

*Ascorbate Peroxidase* (*APX*) is a potential participant in the ascorbate-glutathione cycle (Gill et al. 2012), and is located mainly in the cytoplasm, chloroplasts and other cellular organelles where APX along with gutathione reductase (GR) control cellular redox status. Ascorbic peroxidase requires two molecules of ascorbate to reduce $H_2O_2$ to $O_2$ and $H_2O$ (Noctor and Foyer 1998), which once oxidized, is reduced by GSH (generated from oxidized glutathione (GSSG) by GR). Cadmium can activate APX (Nehnevajova et al. 2012; Parlak and Yilmaz 2013; Ali et al. 2013; Ehsan et al. 2014; Saidi et al. 2014).

*Gutathione Reductase* (*GR*) helps in maintaining a high GSH/GSSG ratio (Foyer et al. 1997), which is crucial for protection against oxidative damage. The activation of GR by Cd has been reported by Bashir et al. (2013), Saidi et al. (2014) and Howladar (2014).

*Ascorbic Acid* (*AsA*) is the most abundant antioxidative enzyme in plants and participates in $H_2O_2$ detoxification via the AsA-GSH cycle which involves (among other things) AsA, GSH, APX, and GR (Noctor and Foyer 1998; Chao et al. 2010). AsA is not only involved in $H_2O_2$ detoxification, but also reacts with $O_2^-$, $OH^-$ and lipid hydroperoxides (Saidi et al. 2013). The exogenous application of AsA reduces Cd toxicity as well as abiotic stresses in plants (Jin et al. 2013). Ascorbate oxidase (AAO), present in plant apoplasts, causes the oxidation of AsA to MDHA using oxygen (Chao et al. 2010).

*Peroxidases* (*PODs*) are antioxidant enzymes that catalyze the $H_2O_2$-dependent oxidation of a number of substrates, especially phenolic compounds (Saidi et al. 2013). Cadmium-induced oxidative stress increases POD activity in plants (Gallego et al. 1996; Lee et al. 2010; Ali et al. 2013; Deng et al. 2014; Howladar 2014).

*Dehydroascorbate reductase* (*DHAR*) is a very important antioxidant in plants, which reduces dehydroascorbate to maintain an appropriate level of ascorbate (ASC), using GSH as the reducing substrate (Nehnevajova et al. 2012). However, no gene or cDNA has been cloned for plant DHAR (Urano et al. 2000). The ASC produced by DHAR is used by APX as a substrate for $H_2O_2$ detoxification. DHAR is reported to contribute to Cd detoxification via its role in the ROS scavenging cycle (Chen et al. 2010; Nehnevajova et al. 2012).

## 6.3 Salicylic Acid (SA)

Salicylic acid (SA) is a hormone-like signaling molecule which influences several physiological and developmental processes in plants. Salicylic acid is involved in the regulation of seed germination, plant growth, flowering, fruit yield, and glycolysis (Guo et al. 2009) and is also known to increase and elicit specific responses as a result of Cd and other trace-metal-induced stress (Ahmad et al. 2011; Tao et al. 2013). Salicylic acid has been shown to reduce Cd-induced oxidative stress in plants (Guo et al. 2007; Tamás et al. 2015) and thereby increase Cd tolerance (Metwally et al. 2003). The protective role of SA primarily includes the induction of gene expression, regulation of antioxidants and ROS, membrane stabilization, and absorption/distribution of essential nutrients (Belkadhi et al. 2015). Plant SA content has been shown to increase in a wide range of species after exposure to Cd and other heavy metals (Tao et al. 2013; Kovács et al. 2014). While the means by which this endogenous signalling molecule is synthesized and regulated in response to abiotic stress (especially in crop plants) requires further study, Dempsey et al. (2011) proposed that SA is synthesized via two distinct pathways: (1) compartmentalized enzymatic pathways that use the metabolite chorismate, and (2) a phenyl-propanoid pathway that uses L-phenylalanine, and synthesizes SA via cinnamate.

Exogenous applications of SA have been shown to mitigate decreases antioxidant enzyme activity and reduce Cd-induced production of TBARS, $H_2O_2$ and $O_2^-$ and lipid peroxidation in *O. sativa* (Panda and Patra 2007). Belkadhi et al. (2013) showed that exogenous SA protected membranes against a loss of integrity and lipid peroxidation by increasing fatty acid unsaturation in *Linum usitatissimum* roots. Later, Belkadhi et al. 2015 further reported that exogenous SA also protected phospholipids of *L. usitatissimum* by decreasing the percentages of monogalactosyldiacylglycerol, phosphatidylcholine, phosphatidylethanolamine and phosphatidylglycerol. The net result of exogenous SA application is the alleviation of Cd-induced root growth inhibition and overall attenuation of Cd toxicity—the latter of which has been reported for a variety of plants (e.g., *V. faba* (Popova et al. 2009), *O. sativa* (Guo et al. 2007), *B. juncea* (Ahmad et al. 2011), *G. max* (Drazic and Mihailovic 2005), *L. usitatissimum* (Belkhadi et al. 2010), *H. vulgare* (Tamás et al. 2015), *Cannabis sativa* (Shi et al. 2009), *Z. mays* (Krantev et al. 2008), *T. aestivum* (Agami and Mohamed 2013) and *Pisum sativum* (Popova et al. 2009)). Salicylic acid is reported to reduce Cd toxicity to plants not only at the level of antioxidant defense but also by altering other mechanisms of Cd detoxification which are not yet well known (Gallego et al. 2012).

Contrary to the above, some authors have postulated that SA itself may cause oxidative stress. Pal et al. (2002)) reported that despite decreasing Cd uptake, exogenous SA application enhanced Cd-induced phytotoxicity. Recently, Liu et al. (2011) showed that pretreatment of *Ricinus communis* with SA alone reduced plant biomass, photosynthetic rate, stomatal conductance and intercellular $CO_2$ concentration. They also reported that SA pretreatment under Cd stress caused a significant decrease in plant biomass, photosynthetic rate, stomatal conductance and intercellular $CO_2$ concentration. Endogenous application of SA may act as a signaling molecule to initiate Cd-induced oxidative stress in *Arabidopsis thaliana* (Zawoznik et al. 2007). Collectively, these studies suggest that the presence of SA (endogenous or exogenous applications) in plants may have contrasting effects in alleviating Cd toxicity, likely due in part to plant type, applied level and form of Cd, and concentration of SA applied.

## 7 Hormetic Effect of Cadmium Toxicity

Cadmium can have adverse effects on the morphological, physiological, and biochemical development and function in plants. The adverse effects of Cd on plants over the complete dose-response range tend to be linearly related to Cd exopsure (Gill et al. 2012; Hayat et al. 2012; Irfan et al. 2014). However, some studies have reported a hormetic effect whereby Cd is toxic at higher applied levels but acts as a growth stimulator at low levels. This biphasic dose–response phenomenon is often described by an inverted U-shaped curve (Jia et al. 2013). Hormetic U-shaped growth effects have frequently been reported for plants exposed to low doses of toxic metal ions such as Cd, Cr, Al, and Pb (Poschenrieder et al. 2013; Calabrese and Baldwin 1999; Calabrese and Baldwin 2003a, b), and have also been reported for malondialdehyde and electrolyte leakage under low Cd exposure. The hormetic effect is not necessarily seen in studies focusing on more highly toxic levels of Cd.

Based on the current knowledge of Cd toxicity and detoxification in plants, hormetic effects of Cd may in part be attributed to adaptive compensatory processes and/or defence mechanisms (Jia et al. 2013; Tamás et al. 2015). This implies the existence of metal-sensing systems that regulate gene expression via cellular signaling transduction cascades. Growth stimulation in plants due to ROS-induced programmed cell death is an important mechanism in developmental processes (Gill and Tuteja 2010). Low level Cd exposure to *T. aestivum* has been shown to prompt a hormetic effect by stimulating plant growth (Lin et al. 2007a, b) which is accompanied by a decrease in ROS production and increase in glutathione reductase activity and GSSG/GSH ratio. An increase in SA concentration (which can similarly stimulate growth) has also been reported in plants when exposed to low levels of Cd (Keunen et al. 2013).

Although perhaps less common, defense or growth-stimulating effects may also be observed in response to higher Cd doses—occurring, for example, in plants pre-acclimated to Cd exposure. The hormetic effect of Cd may also vary with plant species, growth stage, exposure duration, and the time taken by the plant to activate the full suite of metal tolerance mechanisms (Küpper et al. 2007). However, the majority of studies did not investigate the physiological and molecular mechanisms responsible for Cd-induced plant growth stimulation. This is especially the case in studies of Cd toxicity or hyperaccumulation, which instead focused on (1) soil-plant transfer of Cd, (2) accumulation of Cd in different plant tissues, and (3) Cd-induced toxicity to basic growth parameters. For this reason, the response mechanisms of plants subjected to low levels of Cd stress requires further evaluation.

# 8 Conclusions and Perspectives

This review highlighted the biogeochemical behavior of Cd in soil-plant systems. Cadmium adsorption/desorption in soil, mobility towards plant roots and phyto-uptake varies greatly according to the physico-chemical environment. Factors which influence Cd uptake include (but are not limited to) Cd exposure concentration and chemical speciation, soil properties (which influence Cd lability and bioavailability), exposure duration, and plant characteristics (e.g., species type, life stage, and overall fitness). Cadmium uptake by plants, which constitutes the first step of its entrance into the terrestrial food chain, occurs via specific and non-specific transporters of essential nutrients. Inside the plant, Cd mainly accumulates in root cells and a small proportion is translocated to the shoots. Limited translocation of Cd to shoot tissues is due to the blockage of Cd by the Casparian band within the endodermis; that which does make it through is translocated to shoot tissues mainly via different membrane transporter gene families.

Excessive Cd accumulation in plant tissue induces toxicity, which may be manifested as visual impairments such as seed development and germination, root elongation, leaf chlorosis and necrosis, and stunted growth. Excessive Cd accumulation also impairs mineral nutrition and water balance, lamellar organization and chlorophyll production in the chloroplast, cell division, and membrane permeability and structure (by reacting with functional groups of different enzymes). Moreover, Cd inhibits the activity of many enzymes by replacing essential ions and reacting with the active sites of ATP or ADP. Cadmium-induced overproduction of ROS causes DNA damage and lipid peroxidation.

Plants have well-organized enzymatic and non-enzymatic antioxidant defense systems (i.e., defence-related genes and metabolites) which are elicited in response to Cd exposure and collectively used to mitigate toxicity. Plant compartmentalization and detoxification responses include Cd retention in the cell wall (as the first barrier against Cd stress), Cd chelation by PC and subsequent compartmentalization of the PC-Cd complex, and the induction of stress signaling molecules such as salicylic acid. The plasticity of induction of a specific defense mechanism by plants and their efficiency depends on plant species (sensitive or tolerant), exposed Cd level and exposure duration. Because Cd has various negative impacts on environmental sustainability, we believe it deserves to be evaluated in supplementary studies in order to strengthen our understanding about its behavior in soil–plant systems.

## References


Abbas G, Saqib M, Akhtar J, Murtaza G, Shahid M (2015) Effect of salinity on rhizosphere acidification and antioxidant activity of two acacia species. Can J For Res 45:124–129

Agami RA, Mohamed GF (2013) Exogenous treatment with indole-3-acetic acid and salicylic acid alleviates cadmium toxicity in wheat seedlings. Ecotoxicol Environ Saf 94:164–171

Aghababaei F, Raiesi F (2015) Mycorrhizal fungi and earthworms reduce antioxidant enzyme activities in maize and sunflower plants grown in Cd-polluted soils. Soil Biol Biochem 86:87–97

Agnieszka B, Tomasz C, Jerzy W (2014) Chemical properties and toxicity of soils contaminated by mining activity. Ecotoxicology 23:1234–1244

Ahmad P, Nabi G, Ashraf M (2011) Cadmium-induced oxidative damage in mustard [*Brassica juncea* (L.) Czern. & Coss.] plants can be alleviated by salicylic acid. S Afr J Bot 77:36–44

Ahmad I, Akhtar MJ, Asghar HN, Ghafoor U, Shahid M (2015) Differential effects of plant growth-promoting rhizobacteria on maize growth and cadmium uptake. J Plant Growth Regul. doi:10.1007/s00344-015-9534-5

Akhter F, McGarvey B, Macfie SM (2012) Reduced translocation of cadmium from roots is associated with increased production of phytochelatins and their precursors. J Plant Physiol 169:1821–1829

Akoumianakis KA, Passam HC, Barouchas PE, Moustakas NK (2008) Effect of cadmium on yield and cadmium concentration in the edible tissues of endive (*Cichorium endivia* L.) and rocket (*Eruca sativa* Mill.). J Food Agric Environ 6:206–209

Ali H, Khan E, Sajad MA (2013) Phytoremediation of heavy metals—concepts and applications. Chemosphere 91:869–881

Almås AR, Singh BR (2001) Plant uptake of cadmium-109 and zinc-65 at different temperature and organic matter levels. J Environ Qual 30:869–877

Al-Sabti K, Metcalfe CD (1995) Fish micronuclei for assessing genotoxicity in water. Mutat Res 343:121–135

Andosch A, Affenzeller MJ, Lütz C, Lütz-Meindl U (2012) A freshwater green alga under cadmium stress: ameliorating calcium effects on ultrastructure and photosynthesis in the unicellular model Micrasterias. J Plant Physiol 169:1489–1500

Andresen E, Küpper H (2013) Cadmium toxicity in plants. Met Ions Life Sci 11:395–413

Anjum NA, Ahmad I, Mohmood I, Pacheco M, Duarte AC, Pereira E, Umar S, Ahmad A, Khan NA, Iqbal M et al (2012) Modulation of glutathione and its related enzymes in plants' responses to toxic metals and metalloids—a review. Environ Exp Bot 75:307–324

Antolín MC, Muro I, Sánchez-Díaz M (2010) Application of sewage sludge improves growth, photosynthesis and antioxidant activities of nodulated alfalfa plants under drought conditions. Environ Exp Bot 68:75–82

Antunes PMC, Hale BA (2006) The effect of metal diffusion and supply limitations on conditional stability constants determined for durum wheat roots. Plant and Soil 284:229–241

Antunes PMC, Kreager NJ (2014) Lead toxicity to *Lemna minor* predicted using a metal speciation chemistry approach. Environ Toxicol Chem. doi:10.1002/etc.2688

Aravind P, Prasad MNV, Malec P, Waloszek A, Strzałka K (2009) Zinc protects *Ceratophyllum demersum* L. (free-floating hydrophyte) against reactive oxygen species induced by cadmium. J Trace Elem Med Biol 23:50–60

Ardestani MM, van Gestel CAM (2013) Using a toxicokinetics approach to explain the effect of soil pH on cadmium bioavailability to Folsomia candida. Environ Pollut 180:122–130

Armas T, Pinto AP, Varennes A, Mourato MP, Martins LL, Gonçalves MLS, Mota AM (2014) Comparison of cadmium-induced oxidative stress in *Brassica juncea* in soil and hydroponic cultures. Plant Soil. doi:10.1007/s11104-014-2330-3

Arshad M, Silvestre J, Pinelli E, Kallerhoff J, Kaemmerer M, Tarigo A, Shahid M, Guiresse M, Pradere P, Dumat C (2008) A field study of lead phytoextraction by various scented *Pelargonium* cultivars. Chemosphere 71:2187–2192

Asgher M, Khan NA, Iqbal M, Khan R, Fatma M, Masood A (2014) Ethylene production is associated with alleviation of cadmium-induced oxidative stress by sulfur in mustard types differing in ethylene sensitivity. Ecotoxicol Environ Saf 106:54–61

ATSDR (2012) Agency for Toxic Substance and Disease Registry, U.S. toxicological profile for cadmium. Department of Health and Humans Services, Public Health Service, Centers for Disease Control, Atlanta, Georgia, USA

Austruy A, Shahid M, Xiong T, Castrec M, Payre V, Niazi NK, Sabir M, Dumat C (2014) Mechanisms of metal-phosphates formation in the rhizosphere soils of pea and tomato: environmental and sanitary consequences. J Soil Sediments 14:666–678

Bade R, Oh S, Shin WS (2012) Diffusive gradients in thin films (DGT) for the prediction of bioavailability of heavy metals in contaminated soils to earthworm (*Eisenia foetida*) and oral bioavailable concentrations. Sci Total Environ 416:127–136

Baker NR (1991) A possible role for photosystem II in environmental perturbations of photosynthesis. Physiol Plant 81:563–570

Baker AJM, Reeves RD, Hajar ASM (1994) Heavy metal accumulation and tolerance in British populations of the metallophyte *Thlaspi caerulescens* J. & C. Presl (Brassicaceae). New Phytol 127:61–68

Baldantoni D, Morra L, Zaccardelli M, Alfani A (2016) Cadmium accumulation in leaves of leafy vegetables. Ecotoxicol Environ Saf 123:89–94

Bandyopadhyay A, Mukherjee A (2011) Sensitivity of Allium and Nicotiana in cellular and acellular comet assays to assess differential genotoxicity of direct and indirect acting mutagens. Ecotoxicol Environ Saf 74:860–865

Barber JL, Thomas GO, Kerstiens G, Jones KC (2004) Current issues and uncertainties in the measurement and modelling of air-vegetation exchange and within-plant processing of POPs. Environ Pollut 128:99–138

Barrow NJ (1986) Reaction of anions and cations with variable-charge soils. Adv Agron 38:183–230

Baryla A, Carrier P, Franck F, Coulomb C, Sahut C et al (2001) Leaf chlorosis in oilseed rape plants (*Brassica napus*) grown on cadmium-polluted soil: causes and consequences for photosynthesis and growth. Planta 212:696–709

Bashir H, Ahmad J, Bagheri R, Nauman M, Qureshi MI (2013) Limited sulfur resource forces *Arabidopsis thaliana* to shift towards non-sulfur tolerance under cadmium stress. Environ Exp Bot 94:19–32

Behboodi BS, Samadi L (2004) Detection of apoptotic bodies and oligonucleosomal DNA fragments in cadmium-treated root apical cells of *Allium cepa* Linnaeus. Plant Sci 167:411–416

Belimov AA, Kunakova AM, Safronova VI, Stepanok VV, Iudkin LI, Alekseev IV, Kozhemiakov AP (2004) Employment of associative bacteria for the inoculation of barley plants cultivated in soil contaminated with lead and cadmium. Mikrobiologiia 73:118–125

Belimov AA, Hontzeas N, Safronova VI, Demchinskaya SV, Piluzza G, Bullitta S, Glick BR (2005) Cadmium-tolerant plant growth-promoting bacteria associated with the roots of Indian mustard (*Brassica juncea* L. Czern.). Soil Biol Biochem 37:241–250

Belkadhi A, DeHaro A, Soengas P, Obregón S, Cartea ME, Djebali W, Chaïbi W (2013) Salicylic acid improves root antioxidant defense system and total antioxidant capacities of flax subjected to cadmium. OMICS 17:398–406

Belkadhi A, Haro AD, Obregon S, Chaïbi W, Djebali W (2015) Exogenous salicylic acid protects phospholipids against cadmium stress in flax (*Linum usitatissimum* L.). Ecotoxicol Environ Saf 120:102–109

Belkhadi A, Hediji H, Abbes Z, Nouairi I, Barhoumi Z, Zarrouk M, Chaïbi W, Djebali W (2010) Effects of exogenous salicylic acid pre-treatment on cadmium toxicity and leaf lipid content in *Linum usitatissimum* L. Ecotoxicol Environ Saf 73:1004–1011

Benavides MP, Gallego SM, Tomaro ML (2005) Cadmium toxicity in plants. Braz J Plant Physiol 17:21–34

Bilodeau-Gauthier S, Paré D, Messier C, Bélanger N (2011) Juvenile growth of hybrid poplars on acidic boreal soil determined by environmental effects of soil preparation, vegetation control, and fertilization. Forest Ecol Manag 261:620–629

Blair IA (2001) Lipid hydroperoxide-mediated DNA damage. Exp Gerontol 36:1473–1481

Bolan N, Mahimairaja S, Kunhikrishnan A, Naidu R (2013) Sorption–bioavailability nexus of arsenic and cadmium in variable-charge soils. J Hazard Mater 261:725–732

Bramley RGV (1990) Review: cadmium in New Zealand agriculture. New Zeal J Agr Res 33:505–519

Breckle SW, Kahle H (1991) Geobotanyandecology. Prog Bot 52:391–406

Burau RG, Kaita KY, Inouye TS, Miller M (1973) Chemical analysis of soil samples from the Salinis Valley, California for cadmium, zinc, and phosphate. Report to state water resources control board. University of California, Davis, Davis

Cabaniss SE, Zhou Q, Maurice P, Chin Y-P, Aiken GR (2000) A log-normal distribution model for the molecular weight of aquatic fulvic acids. Environ Sci Technol 34:1103–1109

Calabrese EJ, Baldwin LA (1999) Chemical hormesis: its historical foundations as biological hypothesis. Toxicol Pathol 27:195–216

Calabrese EJ, Baldwin LA (2003a) Toxicology rethinks its central belief: hormesis demands a reappraisal of the way risks are assessed. Nature 421:691–692

Calabrese EJ, Baldwin LA (2003b) The hormetic dose-response model is more common than the threshold model in toxicology. Toxicol Sci 71:246–250

Calace N, Petronio BM (2004) The role of organic matter on metal toxicity and bio-availability. Ann Chim 94:487–493

Calderón-Preciado D, Matamoros V, Biel C, Save R, Bayona JM (2013) Foliar sorption of emerging and priority contaminants under controlled conditions. J Hazard Mater 260:176–182

Castillo-Michel HA, Hernandez N, Martinez-Martinez A, Parsons JG, Peralta-Videa JR, Gardea-Torresdey JL (2009) Coordination and speciation of cadmium in corn seedlings and its effects on macro- and micronutrients uptake. Plant Physiol Biochem 47:608–614

Chamel A, Gaillardon P, Gauvrit C (1991) La pénétration foliaire des herbicides. In: Les herbicides: mode d'action et principes d'utilisation. Institut National de la Recherche Agronomique 75007, Paris, pp 8–49

Chaney RL (2012) Chapter two—food safety issues for mineral and organic fertilizers. In : Donald L. Sparks (ed) Adv Agron. Academic, pp 51–116

Chao Y-Y, Hong C-Y, Kao CH (2010) The decline in ascorbic acid content is associated with cadmium toxicity of rice seedlings. Plant Physiol Biochem 48:374–381

Chavez E, He ZL, Stoffella PJ, Mylavarapu RS, Li YC, Moyano B, Baligar VC (2015) Concentration of cadmium in cacao beans and its relationship with soil cadmium in southern Ecuador. Sci Total Environ 533:205–214

Chekmeneva E, Gusmão R, Díaz-Cruz JM, Ariño C, Esteban M (2011) From cysteine to longer chain thiols: thermodynamic analysis of cadmium binding by phytochelatins and their fragments. Metallomics 3:838–846

Chen L, Guo Y, Yang L, Wang Q (2007) SEC-ICP-MS and ESI-MS/MS for analyzing in vitro and in vivo Cd-phytochelatin complexes in a Cd-hyperaccumulator *Brassica chinensis*. J Anal At Spectrom 22:1403–1408

Chen F, Wang F, Wu F, Mao W, Zhang G, Zhou M (2010) Modulation of exogenous glutathione in antioxidant defense system against Cd stress in the two barley genotypes differing in Cd tolerance. Plant Physiol Biochem 48:663–672

Chen JH, Ni JC, Liu QL, Li SX (2012) Adsorption behavior of Cd(II) ions on humic acid-immobilized sodium alginate and hydroxyl ethyl cellulose blending porous composite membrane adsorbent. Desalination 285:54–61

Chen C, Zhou Q, Cai Z (2014a) Effect of soil HHCB on cadmium accumulation and phytotoxicity in wheat seedlings. Ecotoxicology 23:1996–2004

Chen Y, Wu P, Shao Y, Ying Y (2014b) Health risk assessment of heavy metals in vegetables grown around battery production area. Sci Agric 71:126–132

Cherif J, Mediouni C, Ben Ammar W, Jemal F (2011) Interactions of zinc and cadmium toxicity in their effects on growth and in antioxidative systems in tomato plants (*Solanum lycopersicum*). J Environ Sci (China) 23:837–844

Chmielowska-Bąk J, Gzyl J, Rucińska-Sobkowiak R, Arasimowicz-Jelonek M, Deckert J (2014) The new insights into cadmium sensing. Front Plant Sci 5:245

Choppala G, Bolan N, Bibi S, Iqbal M, Rengel Z, Kunhikrishnan A, Ashwath N, Ok YS (2014) Cellular mechanisms in higher plants governing tolerance to cadmium toxicity. Crit Rev Plant Sci 33:374–391

Chou T-S, Chao Y-Y, Kao CH (2012) Involvement of hydrogen peroxide in heat shock- and cadmium-induced expression of ascorbate peroxidase and glutathione reductase in leaves of rice seedlings. J Plant Physiol 169:478–486

Circu ML, Aw TY (2010) Reactive oxygen species, cellular redox systems, and apoptosis. Free Radic Biol Med 48:749–762

Clemens S, Antosiewicz DM, Ward JM, Schachtman DP, Schroeder JI (1998) The plant cDNA LCT1 mediates the uptake of calcium and cadmium in yeast. Proc Natl Acad Sci U S A 95:12043–12048

Clemens S, Aarts MGM, Thomine S, Verbruggen N (2013) The key to preventing slow cadmium poisoning. Trends Plant Sci 18:92–99

Cobbett CS (2000) Phytochelatins and their roles in heavy metal detoxification. Plant Physiol 123:825–832

Cobelo-García A, Santos-Echeandía J, Prego R, Nieto O (2005) Direct simultaneous determination of Cu, Ni and V in seawater using adsorptive cathodic stripping voltammetry with mixed ligands. Electroanalysis 17:906–911

Cohen CK, Fox TC, Garvin DF, Kochian LV (1998) The role of iron-deficiency stress responses in stimulating heavy-metal transport in plants. Plant Physiol 116:1063–1072

Corguinha APB, Gonçalves VC, de Souza GA, de Lima WEA, Penido ES, Pinto CABP, Francisco EAB, Guilherme LRG (2012) Cadmium in potato and soybeans: do phosphate fertilization and soil management systems play a role? J Food Comp Anal 27:32–37

Cortés-Martínez R, Martínez-Miranda V, Solache-Ríos M, García-Sosa I (2004) Evaluation of natural and surfactant-modified zeolites in the removal of cadmium from aqueous solutions. Sep Sci Technol 39(11):2711–2730

Curie C, Cassin G, Couch D, Divol F, Higuchi K, Le Jean M, Misson J, Schikora A, Czernic P, Mari S (2009) Metal movement within the plant: contribution of nicotianamine and yellow stripe 1-like transporters. Ann Bot 103:1–11

Cuypers A, Smeets K, Ruytinx J, Opdenakker K, Keunen E, Remans T, Horemans N, Vanhoudt N, Van Sanden S, Van Belleghem F et al (2011) The cellular redox state as a modulator in cadmium and copper responses in *Arabidopsis thaliana* seedlings. J Plant Physiol 168:309–316

Dabrin A, Durand CL, Garric J, Geffard O, Ferrari BJD, Coquery M (2012) Coupling geochemical and biological approaches to assess the availability of cadmium in freshwater sediment. Sci Total Environ 424:308–315

Dahmani-Muller H, van Oort F, Balabane M (2001) Metal extraction by *Arabidopsis halleri* grown on an unpolluted soil amended with various metal-bearing solids: a pot experiment. Environ Pollut 114:77–84

Dary M, Chamber-Pérez MA, Palomares AJ, Pajuelo E (2010) "In situ" phytostabilisation of heavy metal polluted soils using *Lupinus luteus* inoculated with metal resistant plant-growth promoting rhizobacteria. J Hazard Mater 177:323–330

Das P, Samantaray S, Rout GR (1998) Studies on cadmium toxicity in plants: a review. Environ Pollut 98:29–36

Daud MK, Quiling H, Lei M, Ali B, Zhu SJ (2015) Ultrastructural, metabolic and proteomic changes in leaves of upland cotton in response to cadmium stress. Chemosphere 120:309–320

Degryse F, Shahbazi A, Verheyen L, Smolders E (2012) Diffusion limitations in root uptake of cadmium and zinc, but not nickel, and resulting bias in the Michaelis constant. Plant Physiol 160:1097–1109

Del Río LA (2011) Peroxisomes as a cellular source of reactive nitrogen species signal molecules. Arch Biochem Biophys 506:1–11

Delmail D, Labrousse P, Hourdin P, Larcher L, Moesch C, Botineau M (2011) Physiological, anatomical and phenotypical effects of a cadmium stress in different-aged chlorophyllian organs of *Myriophyllum alterniflorum* DC (Haloragaceae). Environ Exp Bot 72:174–181

Dempsey DMA, Vlot AC, Wildermuth MC, Klessiga DF (2011) Salicylic acid biosynthesis and metabolism. Arabidopsis Book 9:1–156

Deng X, Xia Y, Hu W, Zhang H, Shen Z (2010) Cadmium-induced oxidative damage and protective effects of N-acetyl-L-cysteine against cadmium toxicity in *Solanum nigrum* L. J Hazard Mater 180:722–729

Deng G, Li M, Li H, Yin L, Li W (2014) Exposure to cadmium causes declines in growth and photosynthesis in the endangered aquatic fern (*Ceratopteris pteridoides*). Aquat Bot 112:23–32

Dguimi MH, Debouba M, Ghorbel MH, Gouia H (2009) Tissue-specific cadmium accumulation and its effects on nitrogen metabolism in tobacco (*Nicotiana tabaccum*, Bureley v. Fb9). C R Biol 332:58–68

Dong J, Wu F, Zhang G (2006) Influence of cadmium on antioxidant capacity and four microelement concentrations in tomato seedlings (*Lycopersicon esculentum*). Chemosphere 64:1659–1666

Douay F, Pruvot C, Waterlot C, Fritsch C, Fourrier H, Loriette A, Bidar G, Grand C, de Vaufleury A, Scheifler R (2009) Contamination of woody habitat soils around a former lead smelter in the North of France. Sci Total Environ 407:5564–5577

Douchiche O, Chaïbi W, Morvan C (2012) Cadmium tolerance and accumulation characteristics of mature flax, cv. Hermes: contribution of the basal stem compared to the root. J Hazard Mater 235–236:101–107

Drazic G, Mihailovic N (2005) Modification of cadmium toxicity in soybean seedlings by salicylic acid. Plant Sci 168:511–517

Dumat C, Quenea K, Bermond A, Toinen S, Benedetti MF (2006) Study of the trace metal ion influence on the turnover of soil organic matter in cultivated contaminated soils. Environ Pollut 142:521–529

Duplay J, Semhi K, Errais E, Imfeld G, Babcsanyi I, Perrone T (2014) Copper, zinc, lead and cadmium bioavailability and retention in vineyard soils (Rouffach, France): the impact of cultural practices. Geoderma 230–231:318–328

Duponnois R, Kisa M, Assigbetse K, Prin Y, Thioulouse J, Issartel M, Moulin P, Lepage M (2006) Fluorescent pseudomonads occuring in *Macrotermes subhyalinus* mound structures decrease Cd toxicity and improve its accumulation in sorghum plants. Sci Total Environ 370:391–400

Ďurčeková K, Huttová J, Mistrík I, Ollé M, Tamás L (2007) Cadmium induces premature xylogenesis in barley roots. Plant and Soil 290:61–68

Ebbs S, Lau I, Ahner B, Kochian L (2002) Phytochelatin synthesis is not responsible for Cd tolerance in the Zn/Cd hyperaccumulator *Thlaspi caerulescens* (J. & C. Presl). Planta 214:635–640

Egle K, Römer W, Keller H (2003) Exudation of low molecular weight organic acids by *Lupinus albus* L., *Lupinus angustifolius* L. and *Lupinus luteus* L. as affected by phosphorus supply. Agronomie 23:511–518

Ehsan S, Ali S, Noureen S, Mahmood K, Farid M, Ishaque W, Shakoor MB, Rizwan M (2014) Citric acid assisted phytoremediation of cadmium by *Brassica napus* L. Ecotoxicol Environ Saf 106:164–172

Eikmann T, Kloke A, Eikmann S (1993) Environmental medical and toxicological assessment of soil contamination. In: Arendt F, Annokkée GJ, Bosman R, Brink WJVD (eds) Contaminated soil'93. Springer, Netherlands, pp 327–336

El-Boshy ME, Risha EF, Abdelhamid FM, Mubarak MS, Hadda TB (2014) Protective effects of selenium against cadmium induced hematological disturbances, immunosuppressive, oxidative stress and hepatorenal damage in rats. J Trace Elem Med Biol. doi:10.1016/j.jtemb.2014.05.009

Emsley J (2011) Nature's building blocks: an A-Z guide to the elements. Oxford University Press, Oxford

Epelde L, Becerril JM, Barrutia O, González-Oreja JA, Garbisu C (2010) Interactions between plant and rhizosphere microbial communities in a metalliferous soil. Environ Pollut 158:1576–1583

Fang Y, Sun X, Yang W, Ma N, Xin Z, Fu J, Liu X, Liu M, Mariga AM, Zhu X et al (2014) Concentrations and health risks of lead, cadmium, arsenic, and mercury in rice and edible mushrooms in China. Food Chem 147:147–151

FAO/WHO (1978) Joint FAO/WHO Expert Committee on food additives. Technical report series. 631. Geneva

FAO/WHO (2006) Joint FAO/WHO Food Standard Programme, codex committee on food additives and contaminants, 38th session. The Netherlands

Farooq MA, Ali S, Hameed A, Ishaque W, Mahmood K, Iqbal Z (2013) Alleviation of cadmium toxicity by silicon is related to elevated photosynthesis, antioxidant enzymes; suppressed cadmium uptake and oxidative stress in cotton. Ecotoxicol Environ Saf 96:242–249

Fässler E, Plaza S, Pairraud A, Gupta SK, Robinson B, Schulin R (2011) Expression of selected genes involved in cadmium detoxification in tobacco plants grown on a sulphur-amended metal-contaminated field. Environ Exp Bot 70:158–165

Feng Z, Hu W, Amin S, Tang MS (2003) Mutational spectrum and genotoxicity of the major lipid peroxidation product, trans-4-hydroxy-2-nonenal, induced DNA adducts in nucleotide excision repair-proficient and -deficient human cells. Biochemistry 42:7848–7854

Feng-tao L, Jian-min Q, Gao-yang Z, Li-hui L, Ping-ping F, Ai-fen T, Jian-tang X (2013) Effect of cadmium stress on the growth, antioxidative enzymes and lipid peroxidation in two kenaf (*Hibiscus cannabinus* L.) plant seedlings. J Integr Agri 12:610–620

Fidalgo F, Freitas R, Ferreira R, Pessoa AM, Teixeira J (2011) *Solanum nigrum* L. antioxidant defence system isozymes are regulated transcriptionally and posttranslationally in Cd-induced stress. Environ Exp Bot 72:312–319

Filipič M (2012) Mechanisms of cadmium induced genomic instability. Mutat Res 733:69–77

Flores-Caceres ML, Hattab S, Hattab S, Boussetta H, Banni M, Hernndez LE (2015) Specific mechanisms of tolerance to copper and cadmium are compromised by a limited concentration of glutathione in alfalfa plants. Plant Sci 233:165–173

Fodor E, Szabó-Nagy A, Erdei L (1995) The effects of cadmium on the fluidity and H+-ATPase activity of plasma membrane from sunflower and wheat roots. J Plant Physiol 147(1):87–92

Foltête A-S, Masfaraud J-F, Férard J-F, Cotelle S (2012) Is there a relationship between early genotoxicity and life-history traits in *Vicia faba* exposed to cadmium-spiked soils? Mutat Res 747:159–163

Foucault Y, Lévêque T, Xiong T, Schreck E, Austruy A, Shahid M, Dumat C (2013) Green manure plants for remediation of soils polluted by metals and metalloids: ecotoxicity and human bioavailability assessment. Chemosphere 93:1430–1435

Foyer CH, Lopez-Delgado H, Dat JF, Scott IM (1997) Hydrogen peroxide- and glutathione-associated mechanisms of acclimatory stress tolerance and signalling. Physiol Plant 100:241–254

Fritsch C, Giraudoux P, Cœurdassier M, Douay F, Raoul F, Pruvot C, Waterlot C, de Vaufleury A, Scheifler R (2010) Spatial distribution of metals in smelter-impacted soils of woody habitats: influence of landscape and soil properties, and risk for wildlife. Chemosphere 81:141–155

Fujimaki S, Suzui N, Ishioka NS, Kawachi N, Ito S, Chino M, Nakamura S (2010) Tracing cadmium from culture to spikelet: noninvasive imaging and quantitative characterization of absorption, transport, and accumulation of cadmium in an intact rice plant1[W][OA]. Plant Physiol 152:1796–1806

Gad SC (2014) Cadmium. In: Wexler P (ed) Encycloped toxicol, 3rd edn. Academic, Oxford, pp 613–616

Gallego SM, Benavídes MP, Tomaro ML (1996) Effect of heavy metal ion excess on sunflower leaves: evidence for involvement of oxidative stress. Plant Sci 121:151–159

Gallego SM, Pena LB, Barcia RA, Azpilicueta CE, Iannone MF, Rosales EP, Zawoznik MS, Groppa MD, Benavides MP (2012) Unravelling cadmium toxicity and tolerance in plants: insight into regulatory mechanisms. Environ Exp Bot 83:33–46

Gao Z, Bandosz TJ, Zhao Z, Han M, Qiu J (2009) Investigation of factors affecting adsorption of transition metals on oxidized carbon nanotubes. J Hazard Mater 167:357–365

Garate A, Ramos I, Manzanares M, Lucena JJ (1993) Cadmium uptake and distribution in three cultivars of Lactuca sp. Bull Environ Contam Toxicol 50:709–716

Garnier L, Simon-Plas F, Thuleau P, Agnel J-P, Blein J-P, Ranjeva R, Montillet J-L (2006) Cadmium affects tobacco cells by a series of three waves of reactive oxygen species that contribute to cytotoxicity. Plant Cell Environ 29:1956–1969

Gill SS, Tuteja N (2010) Reactive oxygen species and antioxidant machinery in abiotic stress tolerance in crop plants. Plant Physiol Biochem 48:909–930

Gill SS, Khan NA, Tuteja N (2012) Cadmium at high dose perturbs growth, photosynthesis and nitrogen metabolism while at low dose it up regulates sulfur assimilation and antioxidant machinery in garden cress (*Lepidium sativum* L.). Plant Sci 182:112–120

Gill SS, Hasanuzzaman M, Nahar K, Macovei A, Tuteja N (2013) Importance of nitric oxide in cadmium stress tolerance in crop plants. Plant Physiol Biochem 63:254–261

Goix S, Lévêque T, Xiong T-T, Schreck E, Baeza-Squiban A, Geret F, Uzu G, Austruy A, Dumat C (2014) Environmental and health impacts of fine and ultrafine metallic particles: assessment of threat scores. Environ Res 133:185–194

Gonçalves JF, Becker AG, Cargnelutti D, Tabaldi LA, Pereira LB, Battisti V, Spanevello RM, Morsch VM, Nicoloso FT, Schetinger MRC (2007) Cadmium toxicity causes oxidative stress and induces response of the antioxidant system in cucumber seedlings. Braz J Plant Physiol 19:223–232

Gonçalves JF, Antes FG, Maldaner J, Pereira LB, Tabaldi LA, Rauber R, Rossato LV, Bisognin DA, Dressler VL, Flores EM et al (2009) Cadmium and mineral nutrient accumulation in potato plantlets grown under cadmium stress in two different experimental culture conditions. Plant Physiol Biochem 47:814–821

Gonneau C, Genevois N, Frérot H, Sirguey C, Sterckeman T (2014) Variation of trace metal accumulation, major nutrient uptake and growth parameters and their correlations in 22 populations of *Noccaea caerulescens*. Plant and Soil 384:271–287

Gratão PL, Monteiro CC, Rossi ML, Martinelli AP, Peres LEP, Medici LO, Lea PJ, Azevedo RA (2009) Differential ultrastructural changes in tomato hormonal mutants exposed to cadmium. Environ Exp Bot 67:387–394

Groppa MD, Rosales EP, Iannone MF, Benavides MP (2008) Nitric oxide, polyamines and Cd-induced phytotoxicity in wheat roots. Phytochemistry 69:2609–2615

Gu C-S, Liu L, Zhao Y-H, Deng Y, Zhu X, Huang S-Z (2014) Overexpression of *Iris. lactea* var. *chinensis* metallothionein IlMT2a enhances cadmium tolerance in *Arabidopsis thaliana*. Ecotoxicol Environ Saf 105:22–28

Guan Z, Chai T, Zhang Y, Xu J, Wei W (2009) Enhancement of Cd tolerance in transgenic tobacco plants overexpressing a Cd-induced catalase cDNA. Chemosphere 76:623–630

Guerrero NRV, Nahabedian DE, Wider EA (2000) Analysis of some factors that may modify the bioavailability of cadmium and lead by *Biomphalaria glabrata*. Environ Toxicol Chem 19 (11):2762–2768

Guo B, Liang YC, Zhu YG, Zhao FJ (2007) Role of salicylic acid in alleviating oxidative damage in rice roots (*Oryza sativa*) subjected to cadmium stress. Environ Pollut 147:743–749

Guo B, Liang Y, Zhu Y (2009) Does salicylic acid regulate antioxidant defense system, cell death, cadmium uptake and partitioning to acquire cadmium tolerance in rice? J Plant Physiol 166:20–31

Guo H, Tian R, Zhu J, Zhou H, Pei D, Wang X (2012) Combined cadmium and elevated ozone affect concentrations of cadmium and antioxidant systems in wheat under fully open-air conditions. J Hazard Mater 209–210:27–33

Gustafsson JP (2008) Visual MINTEQ. Version 2.60. Stockholm. http://www.lwr.kth.se/English/OurSoftware/vminteq/

Hagin N, Kono M (1955) A study on the cause of Itai-itai disease. In: Proceedings of the17th meeting of the Japanese Society of Clinical Surgeons (in Japanese)

Hanikenne M, Talke IN, Haydon MJ, Lanz C, Nolte A, Motte P, Kroymann J, Weigel D, Krämer U (2008) Evolution of metal hyperaccumulation required cis-regulatory changes and triplication of HMA4. Nature 453:391–395

Harrison RM, Chirgawi MB (1989) The assessment of air and soil as contributors of some trace metals to vegetable plants I: use of a filtered air growth cabinet. Sci Total Environ 83:13–34

Hasan SA, Fariduddin Q, Ali B, Hayat S, Ahmad A (2009) Cadmium: toxicity and tolerance in plants. J Environ Biol 30:165–174

Hasan SA, Hayat S, Ahmad A (2011) Brassinosteroids protect photosynthetic machinery against the cadmium induced oxidative stress in two tomato cultivars. Chemosphere 84:1446–1451

Hayat S, Alyemeni MN, Hasan SA (2012) Foliar spray of brassinosteroid enhances yield and quality of *Solanum lycopersicum* under cadmium stress. Saudi J Biol Sci 19:325–335

Haydon MJ, Cobbett CS (2007) Transporters of ligands for essential metal ions in plants. New Phytol 174:499–506

He HP, Guo JG, Zhu JX, Yang D (2001) An experimental study of adsorption capacity of Montmorillonite, Kaolinite and Illite for heavy metals. Acta Petrologica et Mineralogica 4:042
Hédiji H, Djebali W, Belkadhi A, Cabasson C, Moing A, Rolin D (2015) Impact of long-term cadmium exposure on mineral content of *Solanum lycopersicum* plants: consequences on fruit production. S Afr J Bot 97:176–181
Heinrichs H, Schulz-Dobrick B, Wedepohl KH (1980) Terrestrial geochemistry of Cd, Bi, Tl, Pb, Zn and Rb. Geochim Cosmochim Acta 44:1519–1533
Helios-Rybicka E, Wójcik R (2012) Competitive sorption/desorption of Zn, Cd, Pb, Ni, Cu, and Cr by clay-bearing mining wastes. Appl Clay Sci 65–66:6–13
Herbette S, Taconnat L, Hugouvieux V, Piette L, Magniette MLM, Cuine S, Auroy P, Richaud P, Forestier C, Bourguignon J et al (2006) Genome-wide transcriptome profiling of the early cadmium response of Arabidopsis roots and shoots. Biochimie 88:1751–1765
Hernandez LE, Carpena-Ruiz R, Gárate A (1996) Alterations in the mineral nutrition of pea seedlings exposed to cadmium. J Plant Nutr 19:1581–1598
Heyno E, Klose C, Krieger-Liszkay A (2008) Origin of cadmium-induced reactive oxygen species production: mitochondrial electron transfer versus plasma membrane NADPH oxidase. New Phytol 179:687–699
Hirsch D, Banin A (1990) Cadmium speciation in soil solutions. J Environ Qual 19:366–372
Hodoshima H, Enomoto Y, Shoji K, Shimada H, Goto F, Yoshihara T (2007) Differential regulation of cadmium-inducible expression of iron-deficiency-responsive genes in tobacco and barley. Physiol Plant 129:622–634
Holmgren GGS, Meyer MW, Chaney RL, Daniels RB (1993) Cadmium, lead, zinc, copper, and nickel in agricultural soils of the United States of America. J Environ Qual 22:335–348
Hossain Z, Makino T, Komatsu S (2012) Proteomic study of β-aminobutyric acid-mediated cadmium stress alleviation in soybean. J Proteomics 75:4151–4164
Hovmand MF, Tjell JC, Mosbaek H (1983) Plant uptake of airborne cadmium. Environ Pollut Ser A Ecol Biol 30:27–38
Howladar M (2014) A novel *Moringa oleifera* leaf extractcan mitigate the stress effects of salinity and cadmium in bean (*Phaseolus vulgaris* L.) plants seed. Ecotoxicol Environ Saf 100:69–75
Hu Y, Norton GJ, Duan G, Huang Y, Liu Y (2014) Effect of selenium fertilization on the accumulation of cadmium and lead in rice plants. Plant and Soil 384:131–140
Huang J, Zhang Y, Peng JS, Zhong C, Yi HY, Ow DW, Gong JM (2012) Fission yeast HMT1 lowers seed cadmium through phytochelatin-dependent vacuolar sequestration in Arabidopsis. Plant Physiol 158:1779–1788
Huguet S, Bert V, Laboudigue A, Barthès V, Isaure M-P, Llorens I, Schat H, Sarret G (2012) Cd speciation and localization in the hyperaccumulator *Arabidopsis halleri*. Environ Exp Bot 82:54–65
Iannone MF, Rosales EP, Groppa MD, Benavides MP (2010) Reactive oxygen species formation and cell death in catalase-deficient tobacco leaf disks exposed to cadmium. Protoplasma 245:15–27
Iannone MF, Groppa MD, Benavides MP (2015) Cadmium induces different biochemical responses in wild type and catalase-deficient tobacco plants. Environ Exp Bot 109:201–211
Inglot P, Lewinska A, Potocki L, Oklejewicz B, Tabecka-Lonczynska A, Koziorowski M, Bugno-Poniewierska M, Bartosz G, Wnuk M (2012) Cadmium-induced changes in genomic DNA-methylation status increase aneuploidy events in a pig Robertsonian translocation model. Mutat Res 747:182–189
Inouhe M, Ito R, Ito S, Sasada N, Tohoyama H, Joho M (2000) Azuki bean cells are hypersensitive to cadmium and do not synthesize phytochelatins. Plant Physiol 123:1029–1036
Iqbal N, Masood A, Nazar R, Syeed S, Khan NA (2010) Photosynthesis, growth and antioxidant metabolism in mustard (*Brassica juncea* L.) cultivars differing in cadmium tolerance. Agricult Sci China 9:519–527
Irfan M, Ahmad A, Hayat S (2014) Effect of cadmium on the growth and antioxidant enzymes in two varieties of *Brassica juncea*. Saudi J Biol Sci 21:125–131

Isaure M-P, Fayard B, Sarret G, Pairis S, Bourguignon J (2006) Localization and chemical forms of cadmium in plant samples by combining analytical electron microscopy and X-ray spectromicroscopy. Spectrochim Acta Part B 61:1242–1252

Ivanov VV (1996) Ekologicheskaya geokhimia elementov (in Russian). Ekologia, kn. 1–6, Moskva. Environmental Geochemistry of Elements Ecology, 1–6:Moscow

Jakubowska D, Janicka-Russak M, Kabała K, Migocka M, Reda M (2015) Modification of plasma membrane NADPH oxidase activity in cucumber seedling roots in response to cadmium stress. Plant Sci 234:50–59

Järup L, Akesson A (2009) Current status of cadmium as an environmental health problem. Toxicol Appl Pharmacol 238:201–208

Jia L, He X, Chen W, Liu Z, Huang Y, Yu S (2013) Hormesis phenomena under Cd stress in a hyperaccumulator *Lonicera japonica* Thunb. Ecotoxicology 22:476–485

Jiang XJ, Luo YM, Liu Q, Liu SL, Zhao QG (2004) Effects of cadmium on nutrient uptake and translocation by Indian Mustard. Environ Geochem Health 26:319–324

Jiang J, Xu R, Jiang T, Li Z (2012) Immobilization of Cu(II), Pb(II) and Cd(II) by the addition of rice straw derived biochar to a simulated polluted Ultisol. J Hazard Mater 229–230:145–150

Jiao W, Chen W, Chang AC, Page AL (2012) Environmental risks of trace elements associated with long-term phosphate fertilizers applications: a review. Environ Pollut 168:44–53

Jin YH, Clark AB, Slebos RJC, Al-Refai H, Taylor JA, Kunkel TA, Resnick MA, Gordenin DA (2003) Cadmium is a mutagen that acts by inhibiting mismatch repair. Nat Genet 34:326–329

Jin CW, Mao QQ, Luo BF, Lin XY, Du ST (2013) Mutation of mpk6 enhances cadmium tolerance in Arabidopsis plants by alleviating oxidative stress. Plant and Soil 371:387–396

Joubert AVP, Lucas L, Garrido F, Joulian C, Jauzein M (2007) Effect of temperature, gas phase composition, pH and microbial activity on As, Zn, Pb and Cd mobility in selected soils in the Ebro and Meuse Basins in the context of global change. Environ Pollut 148:749–758

Kabata-Pendias A (1993) Behavioural properties of trace metals in soils. Appl Geochem 8, Supplement 2:3–9

Kabata-Pendias A, Sadurski W (2004) Trace elements and compounds in soil. In: Merian E, Anke M, Ihnat M, Stoeppler M (eds) Elements and their compounds in the environment, 2nd edn. Wiley-VCH, Weinheim, pp 79–99

Karagiannidis N, Nikolaou N (2000) Influence of arbuscular mycorrhizae on heavy metal (Pb and Cd) uptake, growth, and chemical composition of *Vitis vinifera* L. (cv. Razaki). Am J Enolog Viticult 51:269–275

Keunen K, Remans T, Opdenakker K, Jozefczak M, Gielen H, Guisez Y, Vangronsveld J, Cuypers A (2013) A mutant of the *Arabidopsis thaliana* LIPOXYGENASE1 gene shows altered signalling and oxidative stress related responses after cadmium exposure. Plant Physiol Biochem 63:272–280

Khan MIR, Nazir F, Asgher M, Per TS, Khan NA (2015) Selenium and sulfur influence ethylene formation and alleviate cadmium-induced oxidative stress by improving proline and glutathione production in wheat. J Plant Physiol 173:9–18

Khoudi H, Maatar Y, Gouiaa S, Masmoudi K (2012) Transgenic tobacco plants expressing ectopically wheat $H^+$-pyrophosphatase ($H^+$-PPase) gene TaVP1 show enhanced accumulation and tolerance to cadmium. J Plant Physiol 169:98–103

Kabata-Pendias (2011) Trace elements in soils and plants. 4th edition, pp 1–505

Kopittke PM, Blamey FPC, Menzies NW (2010) Toxicity of Cd to signal grass (*Brachiaria decumbens* Stapf.) and Rhodes grass (*Chloris gayana* Kunth.). Plant and Soil 330:515–523

Kopyra M, Gwóźdź EA (2003) Nitric oxide stimulates seed germination and counteracts the inhibitory effect of heavy metals and salinity on root growth of *Lupinus luteus*. Plant Physiol Biochem 41:1011–1017

Koren S, Arčon I, Kump P, Nečemer M, Vogel-Mikuš K (2013) Influence of $CdCl_2$ and $CdSO_4$ supplementation on Cd distribution and ligand environment in leaves of the Cd *hyperaccumulator Noccaea* (*Thlaspi*) *praecox*. Plant and Soil 370:125–148

Körpe DA, Aras S (2011) Evaluation of copper-induced stress on eggplant (*Solanum melongena* L.) seedlings at the molecular and population levels by use of various biomarkers. Mutat Res 719:29–34

Koutsogiannaki S, Evangelinos N, Koliakos G, Kaloyianni M (2006) Cytotoxic mechanisms of $Zn^{2+}$ and $Cd^{2+}$ involve $Na^+/H^+$ exchanger (NHE) activation by ROS. Aquat Toxicol 78:315–324

Kovács V, Gondor OK, Szalai G, Darkó E, Majláth I, Janda T, Pál M (2014) Synthesis and role of salicylic acid in wheat varieties with different levels of cadmium tolerance. J Hazard Mater 280:12–19

Kovalchuk I, Titov V, Hohn B, Kovalchuk O (2005) Transcriptome profiling reveals similarities and differences in plant responses to cadmium and lead. Mutat Res 570:149–161

Kranner I, Colville L (2011) Metals and seeds: biochemical and molecular implications and their significance for seed germination. Environ Exp Bot 72:93–105

Krantev A, Yordanova R, Janda T, Szalai G, Popova L (2008) Treatment with salicylic acid decreases the effect of cadmium on photosynthesis in maize plants. J Plant Physiol 165:920–931

Krzesłowska M, Lenartowska M, Samardakiewicz S, Bilski H, Woźny A (2010) Lead deposited in the cell wall of Funaria hygrometrica protonemata is not stable—a remobilization can occur. Environ Pollut 158:325–338

Kumar M, Bijo AJ, Baghel RS, Reddy CRK, Jha B (2012) Selenium and spermine alleviate cadmium induced toxicity in the red seaweed *Gracilaria dura* by regulating antioxidants and DNA methylation. Plant Physiol Biochem 51:129–138

Kung C-P, Wu Y-R, Chuang H (2014) Expression of a dye-decolorizing peroxidase results in hypersensitive response to cadmium stress through reducing the ROS signal in Arabidopsis. Environ Exp Bot 101:47–55

Kuo S, McNeal BL (1984) Effects of pH and phosphate on cadmium sorption by a hydrous ferric oxide. Soil Sci Soc Am J 48:1040–1044

Küpper H, Parameswaran A, Leitenmeier BB, Trtilek M, Setlik I (2007) Cadmium induced inhibition of photosynthesis and long-term acclimation to cadmium stress in the hyperaccumulator *Thlaspi caerulescens*. New Phytol 175:655–674

Landrot G, Tappero R, Webb SM, Sparks DL (2012) Arsenic and chromium speciation in an urban contaminated soil. Chemosphere 88:1196–1201

Lee S, An G (2009) Over-expression of OsIRT1 leads to increased iron and zinc accumulations in rice. Plant Cell Environ 32:408–416

Lee K, Bae DW, Kim SH, Han HJ, Liu X, Park HC, Lim CO, Lee SY, Chung WS (2010) Comparative proteomic analysis of the short-term responses of rice roots and leaves to cadmium. J Plant Physiol 167:161–168

Lefèvre I, Marchal G, Edmond Ghanem M, Correal E, Lutts S (2010) Cadmium has contrasting effects on polyethylene glycol-sensitive and resistant cell lines in the Mediterranean halophyte species *Atriplex halimus* L. J Plant Physiol 167:365–374

Li X, Zhou Q, Wei S, Ren W, Sun X (2011) Adsorption and desorption of carbendazim and cadmium in typical soils in northeastern China as affected by temperature. Geoderma 160:347–354

Li L, Liu X, Peijnenburg WJGM, Zhao J, Chen X, Yu J, Wu H (2012) Pathways of cadmium fluxes in the root of the halophyte Suaeda salsa. Ecotoxicol Environ Saf 75:1–7

Li L, Wu H, van Gestel CA, Peijnenburg WJ, Allen HE (2014) Soil acidification increases metal extractability and bioavailability in old orchard soils of Northeast Jiaodong Peninsula in China. Environ Pollut 188:144–152

Lima AIG, Da Cruz e Silva E, Figueira EMPA (2012) Cd-induced signaling pathways in plants: possible regulation of PC synthase by protein phosphatase 1. Environ Exp Bot 79:31–36

Lin A-J, Zhang X-H, Chen M-M, Cao Q (2007a) Oxidative stress and DNA damages induced by cadmium accumulation. J Environ Sci (China) 19:596–602

Lin R, Wang X, Luo Y, Du W, Guo H, Yin D (2007b) Effects of soil cadmium on growth, oxidative stress and antioxidant system in wheat seedlings (*Triticum aestivum* L.). Chemosphere 69:89–98

Lin L, Zhou W, Dai H, Cao F, Zhang G, Wu F (2012) Selenium reduces cadmium uptake and mitigates cadmium toxicity in rice. J Hazard Mater 235–236:343–351

Liptáková Ľ, Huttová J, Mistrík I, Tamás L (2013) Enhanced lipoxygenase activity is involved in the stress response but not in the harmful lipid peroxidation and cell death of short-term cadmium-treated barley root tip. J Plant Physiol 170:646–652

Liu JG, Liang JS, Li KQ, Zhang ZJ, Yu BY, Lu XL, Yang JC, Zhu QS (2003) Correlations between cadmium and mineral nutrients in absorption and accumulation in various genotypes of rice under cadmium stress. Chemosphere 52:1467–1473

Liu W, Li PJ, Qi XM, Zhou QX, Zheng L, Sun TH, Yang YS (2005) DNA changes in barley (*Hordeum vulgare*) seedlings induced by cadmium pollution using RAPD analysis. Chemosphere 61:158–167

Liu D, Kottke I, Adam D (2007) Localization of cadmium in the root cells of *Allium cepa* by energy dispersive X-ray analysis. Biologia Plantarum 51:363–366

Liu C, Li F, Luo C, Liu L, Wang S, Liu T, Li X (2009a) Foliar application of two silica sols reduced cadmium accumulation in rice grains. J Hazard Mater 161:1466–1472

Liu Z, He X, Chen W, Yuan F, Yan K, Tao D (2009b) Accumulation and tolerance characteristics of cadmium in a potential hyperaccumulator—*Lonicera japonica* Thunb. J Hazard Mater 169:170–175

Liu N, Lin Z-F, Lin G-Z, Song L-Y, Chen S-W, Mo H, Peng C-L (2010) Lead and cadmium induced alterations of cellular functions in leaves of *Alocasia macrorrhiza* L. Schott. Ecotoxicol Environ Saf 73:1238–1245

Liu C, Guo J, Cui Y, Lü T, Zhang X, Shi G (2011) Effects of cadmium and salicylic acid on growth, spectral reflectance and photosynthesis of castor bean seedlings. Plant and Soil 344:131–141

Liu X, Lou C, Xu L, Sun L (2012) Distribution and bioavailability of cadmium in ornithogenic coral-sand sediments of the *Xisha archipelago*, South China Sea. Environ Pollut 168:151–160

Liu L, Gong Z, Zhang Y, Li P (2014a) Growth, cadmium uptake and accumulation of maize (*Zea mays* L.) under the effects of arbuscular mycorrhizal fungi. Ecotoxicology 23:1979–1986

Liu Y, Wu F, Mu Y, Feng C, Fang Y, Chen L, Giesy JP (2014b) Setting water quality criteria in China: approaches for developing species sensitivity distributions for metals and metalloids. Rev Environ Contam Toxicol 230:35–57

Liu B, Huang Q, Cai H, Guo X, Wang T, Gui M (2015) Study of heavy metal concentrations in wild edible mushrooms in Yunnan Province, China. Food Chem 188:294–300

Llugany M, Miralles R, Corrales I, Barceló J, Poschenrieder C (2012) *Cynara cardunculus* a potentially useful plant for remediation of soils polluted with cadmium or arsenic. J Geochem Explor 123:122–127

Loeffler S, Hochberger A, Grill E, Winnacker E-L, Zenk MH (1989) Termination of the phytochelatin synthase reaction through sequestration of heavy metals by the reaction product. FEBS Lett 258:42–46

Logan TJ, Chaney R (1984) Metals. In: Page AL, Gleason TL, Smith JE, Iskandar IK, Sommers LE (eds) Utilization of municipal wastewater and sludge on land. University of California Press, Riverside, CA, p 235

Logan TJ, Miller RH (1983) Background levels of heavy metals in Ohio farm soils. Research circular 275. The Ohio State University Agricultural Research and Development, Wooster, Ohio

Lomaglio T, Rocco M, Trupiano D, De Zio E, Grosso A, Marra M, Delfine S, Chiatante D, Morabito D, Scippa GS (2015) Effect of short-term cadmium stress on Populus nigra L. detached leaves. J Plant Physiol 182:40–48

Lombi E, Tearall KL, Howarth JR, Zhao F-J, Hawkesford MJ, McGrath SP (2002) Influence of iron status on cadmium and zinc uptake by different ecotypes of the hyperaccumulator *Thlaspi caerulescens*. Plant Physiol 128:1359–1367

Long LK, Yao Q, Guo J, Yang RH, Huang YH, Zhu HH (2010) Molecular community analysis of arbuscular mycorrhizal fungi associated with five selected plant species from heavy metal polluted soils. Eur J Soil Biol 46:288–294

Louwagie G, Gay SH, Burrell A (2009) Addressing soil degradation in EU agriculture: relevant processes, practices and policies. EUR 23767 EN

Lozano-Rodriguez E, Hernàndez LE, Bonay P, Carpena-Ruiz RO (1997) Distribution of cadmium in shoot and root tissues1. J Exp Bot 48:123–128

Lugon-Moulin N, Zhang M, Gadani F, Rossi L, Koller D, Krauss M, Wagner GJ (2004) Critical review of the scienceand options for reducing cadmium in tobacco (*Nicotiana tabacum* L.) and other plants. Adv Agron. Academic, pp 111–180

Lushchak VI (2007) Free radical oxidation of proteins and its relationship with functional state of organisms. Biochemistry 72:809–827

Lushchak VI (2011) Adaptive response to oxidative stress: bacteria, fungi, plants and animals. Comp Biochem Physiol Part C Toxicol Pharmacol 153:175–190

Lux A, Martinka M, Vaculík M, White PJ (2011) Root responses to cadmium in the rhizosphere: a review. J Exp Bot 62:21–37

Luxford C, Dean RT, Davies MJ (2002) Induction of DNA damage by oxidised amino acids and proteins. Biogerontology 3:95–102

Lyubenova L, Schröder P (2011) Plants for waste water treatment—effects of heavy metals on the detoxification system of *Typha latifolia*. Biores Technol 102:996–1004

Maestri E, Marmiroli M, Visioli G, Marmiroli N (2010) Metal tolerance and hyperaccumulation: costs and trade-offs between traits and environment. Environ Exp Bot 68:1–13

Mahmood A, Malik RN (2014) Human health risk assessment of heavy metals via consumption of contaminated vegetables collected from different irrigation sources in Lahore, Pakistan. Arab J Chem 7:91–99

Maksimović I, Kastori R, Krstić L, Luković J (2007) Steady presence of cadmium and nickel affects root anatomy, accumulation and distribution of essential ions in maize seedlings. Biologia Plantarum 51:589–592

Marcano L, Carruyo I, Del Campo A, Montiel X (2002) Effect of cadmium on the nucleoli of meristematic cells of onion *Allium cepa* L: an ultrastructural study. Environ Res 88:30–35

Marmiroli M, Pigoni V, Savo-Sardaro ML, Marmiroli N (2014) The effect of silicon on the uptake and translocation of arsenic in tomato (*Solanum lycopersicum* L.). Environ Exp Bot 99:9–17

Márquez-García B, Horemans N, Cuypers A, Guisez Y, Córdoba F (2011) Antioxidants in *Erica andevalensis*: a comparative study between wild plants and cadmium-exposed plants under controlled conditions. Plant Physiol Biochem 49:110–115

Martínez-Peñalver A, Graña E, Reigosa MJ, Sánchez-Moreiras AM (2012) The early response of *Arabidopsis thaliana* to cadmium- and copper-induced stress. Environ Exp Bot 78:1–9

Mathys W (1975) Enzymes of heavy-metal resistant and non-resistant populations of silene cucubalus and their interaction with some heavy metals in vitro and in vivo. Physiol Plant 33 (2):161–165

May MJ, Vernoux T, Leaver C, Montagu MV, Inzé D (1998) Glutathione homeostasis in plants: implications for environmental sensing and plant development. J Exp Bot 49:649–667

McLaughlin MJ, Smolders E, Degryse F, Rietra R (2011) Uptake of metals from soil into vegetables. In: Swartjes FA (ed) Dealing with contaminated sites. Springer, Berlin, pp 325–367

Memon AR, Schröder P (2009) Implications of metal accumulation mechanisms to phytoremediation. Environ Sci Pollut Res Int 16:162–175

Mendoza-Cozatl DG, Zhai Z, Jobe TO, Akmakjian GZ, Song WY, Limbo O, Russell MR, Kozlovskyy VI, Martinoia E, Vatamaniuk OK, Russell P, Schroeder JI (2010) Tonoplast-localized Abc2 transporter mediates phytochelatin accumulation in vacuoles and confers cadmium tolerance. J Biol Chem 285:40416–40426

Mendoza-Cózatl DG, Jobe TO, Hauser F, Schroeder JI (2011) Long-distance transport, vacuolar sequestration, tolerance, and transcriptional responses induced by cadmium and arsenic. Curr Opin Plant Biol 14:554–562

Mengel K, Kirkby EA, Kosegarten H, Appel T (2001) Principles of plant nutrition. Springer Science & Business Media

Mesjasz-Przybyłowics J, Nakonieczny M, Migula P, Augustyniak M, Tarnawska M, Reimold WU, Koeberl C, Przybyłowicz W, Głowacka E (2004) Uptake of cadmium, lead, nickel and zinc from soil and water solutions by the nickel hyperaccumulator *berkheya coddii*. Acta Biol Cracoviensia Ser Bot 46:75–85

Metwally A, Finkemeier I, Georgi M, Dietz K-J (2003) Salicylic acid alleviates the cadmium toxicity in barley seedlings. Plant Physiol 132:272–281

Mishra S, Srivastava S, Tripathi RD, Kumar R, Seth CS, Gupta DK (2006) Lead detoxification by coontail (*Ceratophyllum demersum* L.) involves induction of phytochelatins and antioxidant system in response to its accumulation. Chemosphere 65:1027–1039

Miyadate H, Adachi S, Hiraizumi A, Tezuka K, Nakazawa N, Kawamoto T, Katou K, Kodama I, Sakurai K, Takahashi H et al (2011) OsHMA3, a P1B-type of ATPase affects root-to-shoot cadmium translocation in rice by mediating efflux into vacuoles. New Phytol 189:190–199

Mohamed AA, Castagna A, Ranieri A, Sanità di Toppi L (2012) Cadmium tolerance in *Brassica juncea* roots and shoots is affected by antioxidant status and phytochelatin biosynthesis. Plant Physiol Biochem 57:15–22

Mombo S, Foucault Y, Deola F, Gaillard I, Goix S, Shahid M, Schreck E, Pierart A, Dumat C (2016) Management of human health risk in the context of kitchen gardens polluted by lead and cadmium near a lead recycling company. J Soils Sediments. doi:10.1007/s11368-015-1069-7

Monteiro CC, Carvalho RF, Gratão PL, Carvalho G, Tezotto T, Medici LO, Peres LEP, Azevedo RA (2011) Biochemical responses of the ethylene-insensitive Never ripe tomato mutant subjected to cadmium and sodium stresses. Environ Exp Bot 71:306–320

Monteiro C, Santos C, Pinho S, Oliveira H, Pedrosa T, Dias MC (2012) Cadmium-induced cyto- and genotoxicity are organ-dependent in lettuce. Chem Res Toxicol 25:1423–1434

Morel M, Crouzet J, Gravot A, Auroy P, Leonhardt N, Vavasseur A, Richaud P (2009) AtHMA3, a P1B-ATPase allowing Cd/Zn/Co/Pb vacuolar storage in Arabidopsis. Plant Physiol 149:894–904

Morman SA, Plumlee GS (2013) The role of airborne mineral dusts in human disease. Aeolian Res 9:203–212

Muehe EM, Obst M, Hitchcock A, Tyliszczak T, Behrens S, Schröder C, Byrne JM, Michel FM, Krämer U, Kappler A (2013) Fate of Cd during microbial Fe(III) mineral reduction by a novel and Cd-tolerant geobacter species. Environ Sci Technol 47:14099–14109

Munshower FF (1977) Cadmium accumulation in plants and animals of polluted and non polluted grasslands. J Environ Qual 6:411–413

Najeeb U, Jilani G, Ali S, Sarwar M, Xu L, Zhou W (2011) Insights into cadmium induced physiological and ultra-structural disorders in *Juncus effusus* L. and its remediation through exogenous citric acid. J Hazard Mater 186:565–574

Najmanova J, Neumannova E, Leonhardt T, Zitka O, Kizek R, Macek T, Mackova M, Kotrba P (2012) Cadmium-induced production of phytochelatins and speciation of intracellular cadmium in organs of *Linum usitatissimum* seedlings. Ind Crop Prod 36:536–542

Nakanishi H, Ogawa I, Ishimaru Y, Mori S, Nishizawa NK (2006) Iron deficiency enhances cadmium uptake and translocation mediated by the $Fe^{2+}$ transporters OsIRT1 and OsIRT2 in rice. Soil Sci Plant Nutr 52:464–469

Nedjimi B, Daoud Y (2009) Cadmium accumulation in *Atriplex halimus* subsp. schweinfurthii and its influence on growth, proline, root hydraulic conductivity and nutrient uptake. Flora Morphol Distribut Funct Ecol Plants 204:316–324

Nehnevajova E, Lyubenova L, Herzig R, Schröder P, Schwitzguébel J-P, Schmülling T (2012) Metal accumulation and response of antioxidant enzymes in seedlings and adult sunflower

mutants with improved metal removal traits on a metal-contaminated soil. Environ Exp Bot 76:39–48
Niazi NK, Bishop TFA, Singh B (2011a) Evaluation of spatial variability of soil arsenic adjacent to a disused cattle-dip site, using model-based geostatistics. Environ Sci Technol 45:10463–10470
Niazi NK, Singh B, Shah P (2011b) Arsenic speciation and phytoavailability in contaminated soils using a sequential extraction procedure and XANES spectroscopy. Environ Sci Technol 45:7135–7142
Niazi NK, Singh B, Van Zwieten L, Kachenko AG (2011c) Phytoremediation potential of *Pityrogramma calomelanos* var. austroamericana and *Pteris vittata* L. grown at a highly variable arsenic contaminated site. Int J Phytoremediation 13:912–932
Niazi NK, Singh B, Van Zwieten L, Kachenko AG (2012) Phytoremediation of an arsenic contaminated site using *Pteris vittata* L. and *Pityrogramma calomelanos* var. austroamericana: a long-term study. Environ Sci Pollut Res 19:3506–3515
Nishizono H, Ichikawa H, Suziki S, Ishii F (1987) The role of the root cell wall in the heavy metal tolerance of *Athyrium yokoscense*. Plant and Soil 101:15–20
Noctor G, Foyer CH (1998) Ascorbate and glutathione: keeping active oxygen under control. Annu Rev Plant Physiol Plant Mol Biol 49:249–279
Noctor G, Arisi A-CM, Jouanin L, Foyer CH (1998) Manipulation of glutathione and amino acid biosynthesis in the chloroplast. Plant Physiol 118:471–482
Nordic Council of Ministers (2003) Cadmium review. www.who.int/ifcs/documents/forums/forum5/nmr_cadmium.pdf
Nriagu JO (1989) A global assessment of natural sources of atmospheric trace metals. Nature 338:47–48
Nzengue Y, Candéias M, Sauvaigo S, Douki T, Favier A, Rachidi W, Guiraud P (2015) The toxicity redox mechanisms of cadmium alone or together with copper and zinc homeostasis alteration: its redox biomarkers. J Trace Elem Med Biol 25:171–180
Oda K, Otani M, Uraguchi S, Akihiro T, Fujiwara T (2011) Rice ABCG43 is Cd inducible and confers Cd tolerance on yeast. Biosci Biotechnol Biochem 75:1211–1213
OECD (1994) Risk reduction monograph no. 5: cadmium. OECD environment monograph series no. 104. OECD Environment Directorate, Paris
Ogawa S, Yoshidomi T, Yoshimura E (2011) Cadmium(II)-stimulated enzyme activation of *Arabidopsis thaliana* phytochelatin synthase 1. J Inorg Biochem 105:111–117
Olmos E, Martínez-Solano JR, Piqueras A, Hellín E (2003) Early steps in the oxidative burst induced by cadmium in cultured tobacco cells (BY-2 line). J Exp Bot 54:291–301
Page AL, Chang AC, El-Amamy M (1987) Cadmium levels in soils and crops in the United States. Lead, mercury, cadmium and arsenic in the environment. 119–146
Pal M, Szalai G, Horvath E, Janda T, Paldi E (2002) Effect of salicylic acid during heavy metal stress. Acta Biol Szegediensis 46:119–120
Panda SK, Patra HK (2007) Effect of salicylic acid potentiates cadmium-induced oxidative damage in *Oryza sativa* L. leaves. Acta Physiol Plant 29:567–575
Pandey SP, Somssich IE (2009) The role of WRKY transcription factors in plant immunity. Plant Physiol 150:1648–1655
Park Y, Moon Y, Ryoo J, Kim N, Cho H, Ahn JH (2012) Identification of the minimal region in lipase ABC transporter recognition domain of *Pseudomonas fluorescens* for secretion and fluorescence of green fluorescent protein. Microb Cell Fact 11:11–60
Parkhurst DL, Appelo CAJ (1999) User's guide to PHREEQC (version 2). US Geol Surv Water Resour Inv Rep 99-4259, 312p
Parlak UK, Demirezen Yilmaz D (2013) Ecophysiological tolerance of *Lemna gibba* L. exposed to cadmium. Ecotoxicol Environ Saf 91:79–85
Pena LB, Pasquini LA, Tomaro ML, Gallego SM (2007) 20S proteasome and accumulation of oxidized and ubiquitinated proteins in maize leaves subjected to cadmium stress. Phytochemistry 68:1139–1146

Pence NS, Larsen PB, Ebbs SD, Letham DLD, Lasat MM, Garvin DF, Eide D, Kochian LV (2000) The molecular physiology of heavy metal transport in the Zn/Cd hyperaccumulator *Thlaspi caerulescens*. Proc Natl Acad Sci U S A 97:4956–4960

Pereira BFF, Rozane DE, Araujo SR, Barth G, Queiroz RJB, Nogueira TAR, Moraes MF, Cabral CP, Boaretto AE, Malavolta E (2011) Cadmium availability and accumulation by lettuce and rice. Rev Bras Cienc Solo 35:645–654

Pierce FJ, Dowdy RH, Grigal DF (1982) Concentrations of six trace metals in some major Minnesota soil series. J Environ Qual 11:416–422

Piotrowska-Niczyporuk A, Bajguz A, Zambrzycka E, Godlewska-Zylkiewicz B (2012) Phytohormones as regulators of heavy metal biosorption and toxicity in green alga Chlorella vulgaris (Chlorophyceae). Plant Physiol Biochem 52:52–65

Pizzol M, Bulle C, Thomsen M (2012) Indirect human exposure assessment of airborne lead deposited on soil via a simplified fate and speciation modelling approach. Sci Total Environ 421–422:203–209

Podazza G, Arias M, Prado FE (2012) Cadmium accumulation and strategies to avoid its toxicity in roots of the citrus rootstock Citrumelo. J Hazard Mater 215–216:83–89

Popova LP, Maslenkova LT, Yordanova RY, Ivanova AP, Krantev AP, Szalai G, Janda T (2009) Exogenous treatment with salicylic acid attenuates cadmium toxicity in pea seedlings. Plant Physiol Biochem 47:224–231

Poschenrieder C, Cabot C, Martos S, Gallego B, Barceló J (2013) Do toxic ions induce hormesis in plants? Plant Sci 212:15–25

Pourrut B, Perchet G, Silvestre J, Cecchi M, Guiresse M, Pinelli E (2008) Potential role of NADPH-oxidase in early steps of lead-induced oxidative burst in *Vicia faba* roots. J Plant Physiol 165:571–579

Pourrut B, Shahid M, Dumat C, Winterton P, Pinelli E (2011) Lead uptake, toxicity, and detoxification in plants. Rev Environ Contam Toxicol 213:113–136

Pourrut B, Shahid M, Douay F, Dumat C, Pinelli E (2013) Molecular mechanisms involved in lead uptake, toxicity and detoxification in higher plants. In: Gupta DK, Corpas FJ, Palma JM (eds) Heavy metal stress in plants. Springer, Berlin Heidelberg, pp 121–147

Prasad MNV (2004) Heavy metal stress in plants: from biomolecules to ecosystems. Springer Science & Business Media

Préveral S, Gayet L, Moldes C, Hoffmann J, Mounicou S, Gruet A, Reynaud F, Lobinski R, Verbavatz J-M, Vavasseur A et al (2009) A common highly conserved cadmium detoxification mechanism from bacteria to humans: heavy metal tolerance conferred by the ATP-binding cassette (ABC) transporter SpHMT1 requires glutathione but not metal-chelating phytochelatin peptides. J Biol Chem 284:4936–4943

Qi Z-M, Feng S-Y, Helmers MJ (2012) Modeling cadmium transport in neutral and alkaline soil columns at various depths. Pedosphere 22:273–282

Qian H, Li J, Pan X, Jiang H, Sun L, Fu Z (2010) Photoperiod and temperature influence cadmium's effects on photosynthesis-related gene transcription in *Chlorella vulgaris*. Ecotoxicol Environ Saf 73:1202–1206

Qiu R-L, Zhao X, Tang Y-T, Yu F-M, Hu P-J (2008) Antioxidative response to Cd in a newly discovered cadmium hyperaccumulator, *Arabis paniculata* F. Chemosphere 74:6–12

Quartacci MF, Argilla A, Baker AJM, Navari-Izzo F (2006) Phytoextraction of metals from a multiply contaminated soil by Indian mustard. Chemosphere 63:918–925

Quenea K, Lamy I, Winterton P, Bermond A, Dumat C (2009) Interactions between metals and soil organic matter in various particle size fractions of soil contaminated with waste water. Geoderma 149:217–223

Quezada-hinojosa R, Föllmi KB, Gillet F, Matera V (2015) Cadmium accumulation in six common plant species associated with soils containing high geogenic cadmium concentrations at Le Gurnigel, Swiss Jura Mountains EX. Catena 124:85–96

Rady MM (2011) Effect of 24-epibrassinolide on growth, yield, antioxidant system and cadmium content of bean (*Phaseolus vulgaris* L.) plants under salinity and cadmium stress. Sci Hortic 129:232–237

Rady MM, Hemida KA (2015) Modulation of cadmium toxicity and enhancing cadmium-tolerance in wheat seedlings by exogenous application of polyamines. Ecotoxicol Environ Saf 119:178–185

Ramos J, Clemente MR, Naya L, Loscos J, Perez-Rontome C, Sato S, Tabata S, Becana M (2007) Phytochelatin synthases of the model legume *lotus japonicus*. A small multigene family with differential response to cadmium and alternatively spliced variants. Plant Physiol 143:1110–1118

Ran X, Liu R, Xu S, Bai F, Xu J, Yang Y, Shi J, Wu Z (2014) Assessment of growth rate, chlorophyll a fluorescence, lipid peroxidation and antioxidant enzyme activity in *Aphanizomenon flosaquae*, *Pediastrum simplex* and *Synedra acus* exposed to cadmium. Ecotoxicology. doi:10.1007/s10646-014-1395-3

Rao RAK, Kashifuddin M (2012) Adsorption studies of Cd(II) on ball clay: comparison with other natural clays. Arab J Chem. doi:10.1016/j.arabjc.2012.01.010

Rao KS, Mohapatra M, Anand S, Venkateswarlu P (2010) Review on cadmium removal from aqueous solutions. Int J Eng Sci Technol 2:81–103

Rascio N, Navari-Izzo F (2011) Heavy metal hyperaccumulating plants: how and why do they do it? And what makes them so interesting? Plant Sci 180:169–181

Rea PA (2007) Plant ATP-binding cassette transporters. Annu Rev Plant Biol 58:347–375

Rea PA, Vatamaniuk OK, Rigden DJ (2004) Weeds, worms, and more. papain's long-lost cousin, phytochelatin synthase. Plant Physiol 136:2463–2474

Redondo-Gómez S, Mateos-Naranjo E, Andrades-Moreno L (2010) Accumulation and tolerance characteristics of cadmium in a halophytic Cd-hyperaccumulator, *Arthrocnemum macrostachyum*. J Hazard Mater 184:299–307

Renella G, Adamo P, Bianco MR, Landi L, Violante P, Nannipieri P (2004) Availability and speciation of cadmium added to a calcareous soil under various managements. Eur J Soil Sci 55:123–133

Reuter J, Perdue E (1977) Importance of heavy metal-organic matter interactions in natural waters. Geochim Cosmochim Acta 41:325–334

Rinalducci S, Murgiano L, Zolla L (2008) Redox proteomics: basic principles and future perspectives for the detection of protein oxidation in plants. J Exp Bot 59:3781–3801

Rodrigo A, Avila A, Gómez-Bolea A (1999) Trace metal contents in *Parmelia caperata* (L.) Ach. compared to bulk deposition, throughfall and leaf-wash fluxes in two holm oak forests in Montseny (NE Spain). Atmos Environ 33:359–367

Rodríguez-Serrano M, Romero-Puertas MC, Zabalza A, Corpas FJ, Gómez M, Del Río LA, Sandalio LM (2006) Cadmium effect on oxidative metabolism of pea (*Pisum sativum* L.) roots. Imaging of reactive oxygen species and nitric oxide accumulation in vivo. Plant Cell Environ 29:1532–1544

Rodríguez-Serrano M, Romero-Puertas MC, Pazmiño DM, Testillano PS, Risueño MC, Del Río LA, Sandalio LM (2009) Cellular response of pea plants to cadmium toxicity: cross talk between reactive oxygen species, nitric oxide, and calcium. Plant Physiol 150:229–243

Romero-Puertas MC, Palma JM, Gómez M, Del Río LA, Sandalio LM (2002) Cadmium causes the oxidative modification of proteins in pea plants. Plant Cell Environ 25:677–686

Rosén K, Eriksson J, Vinichuk M (2012) Uptake and translocation of 109Cd and stable Cd within tobacco plants (*Nicotiana sylvestris*). J Environ Radioact 113:16–20

Roth U, von Roepenack-Lahaye E, Clemens S (2006) Proteome changes in *Arabidopsis thaliana* roots upon exposure to $Cd^{2+}$. J Exp Bot 57:4003–4013

Sabir M, Waraich EA, Hakeem KR, Öztürk M, Ahmad HR, Shahid M (2015) Phytoremediation: mechanisms and adaptations. Soil Remed Plants 4:85–105

Saeki K, Kunito T (2012) Influence of chloride ions on cadmium adsorptions by oxides, hydroxides, oxyhydroxides, and phyllosilicates. Appl Clay Sci 62–63:58–62

Saidi I, Ayouni M, Dhieb A, Chtourou Y, Chaïbi W, Djebali W (2013) Oxidative damages induced by short-term exposure to cadmium in bean plants: protective role of salicylic acid. S Afr J Bot 85:32–38

Saidi I, Chtourou Y, Djebali W (2014) Selenium alleviates cadmium toxicity by preventing oxidative stress in sunflower (*Helianthus annuus*) seedlings. J Plant Physiol 171:85–91

Salminen R, Batista MJ, Bidovec M, Demetriades A, De Vivo B, De Vos W, Duris M, Gilucis A, Gregorauskiene V, Halamic J, Heitzmann P, Lima A, Jordan G, Klaver G, Klein P, Lis J, Locutura J, Marsina K, Mazreku A, O'Connor PJ, Olsson SÅ, Ottesen RT, Petersell V, Plant JA, Reeder S, Salpeteur I, Sandström H, Siewers U, Steenfelt A, Tarvainen T (2005) Geochemical Atlas of Europe. Part 1 – Background Information, Methodology and Maps. Geological Survey of Finland, Espoo, Finland

Salt DE, Rauser WE (1995) MgATP-dependent transport of phytochelatins across the tonoplast of oat roots. Plant Physiol 107:1293–1301

Salt DE, Wagner GJ (1993) Cadmium transport across tonoplast of vesicles from oat roots. Evidence for a $Cd^{2+}/H^+$ antiport activity. J Biol Chem 268:12297–12302

Sánchez-Marín P, Santos-Echeandía J, Nieto-Cid M, Álvarez-Salgado XA, Beiras R (2010) Effect of dissolved organic matter (DOM) of contrasting origins on Cu and Pb speciation and toxicity to *Paracentrotus lividus* larvae. Aquat Toxicol 96:90–102

Santos-Echeandía J, Vale C, Caetano M, Pereira P, Prego R (2010) Effect of tidal flooding on metal distribution in pore waters of marsh sediments and its transport to water column (Tagus estuary, Portugal). Mar Environ Res 70(5):358–367

Sarret G, Smits EAHP, Michel HC, Isaure MP, Zhao FJ, Tappero R (2013) Chapter one—use of synchrotron-based techniques to elucidate metal uptake and metabolism in plants. Adv Agron 119:1–82

Sauvé S, Hendershot W, Allen H (2000) Solid-solution partitioning of metals in contaminated soils: dependence on pH, total metal burden, and organic matter. Environ Sci Technol 34:1125–1131

Schreck E, Foucault Y, Sarret G, Sobanska S, Cécillon L, Castrec-Rouelle M, Uzu G, Dumat C (2012a) Metal and metalloid foliar uptake by various plant species exposed to atmospheric industrial fallout: mechanisms involved for lead. Sci Total Environ 427–428:253–262

Schreck E, Bonnard R, Laplanche C, Leveque T, Foucault Y, Dumat C (2012b) DECA: a new model for assessing the foliar uptake of atmospheric lead by vegetation, using *Lactuca sativa* as an example. J Environ Manage 112:233–239

Schreck E, Laplanche C, Le Guédard M, Bessoule J-J, Austruy A, Xiong T, Foucault Y, Dumat C (2013) Influence of fine process particles enriched with metals and metalloids on *Lactuca sativa* L. leaf fatty acid composition following air and/or soil-plant field exposure. Environ Pollut 179:242–249

Schreck E, Dappe V, Sarret G, Sobanska S, Nowak D, Nowak J, Stefaniak EA, Magnin V, Ranieri V, Dumat C (2014) Foliar or root exposures to smelter particles: consequences for lead compartmentalization and speciation in plant leaves. Sci Total Environ 476–477:667–676

Schwartz GG, Reis IM (2000) Is cadmium a cause of human pancreatic cancer? Cancer Epidemiol Biomarkers Prev 9:139–145

Scoullos M, Vonkeman GH, Thornton I, Makuch Z (2001) Mercury-cadmium-lead handbook for sustainable heavy metals policy and regulation: handbook for sustainable heavy metals policy and regulation. Springer Science & Business Media, Berlin

Semane B, Cuypers A, Smeets K, Van Belleghem F, Horemans N, Schat H, Vangronsveld J (2007) Cadmium responses in *Arabidopsis thaliana*: glutathione metabolism and antioxidative defence system. Physiol Plant 129:519–528

Seregin IV, Shpigun LK, Ivanov VB (2004) Distribution and toxic effects of cadmium and lead on maize roots. Russian J Plant Physiol 51:525–533

Sergeant K, Kieffer P, Dommes J, Hausman J-F, Renaut J (2014) Proteomic changes in leaves of poplar exposed to both cadmium and low-temperature. Environ Exp Bot 106:112–123

Shah K, Kumar RG, Verma S, Dubey RS (2001) Effect of cadmium on lipid peroxidation, superoxide anion generation and activities of antioxidant enzymes in growing rice seedlings. Plant Sci 161:1135–1144
Shahid M, Pinelli E, Pourrut B, Silvestre J, Dumat C (2011) Lead-induced genotoxicity to Vicia faba L. roots in relation with metal cell uptake and initial speciation. Ecotoxicol Environ Saf 74:78–84
Shahid M, Arshad M, Kaemmerer M, Pinelli E, Probst A, Baque D, Pradere P, Dumat C (2012a) Long-term field metal extraction by Pelargonium: phytoextraction efficiency in relation to plant maturity. Int J Phytoremediation 14:493–505
Shahid M, Dumat C, Aslam M, Pinelli E (2012b) Assessment of lead speciation by organic ligands using speciation models. Chem Spec Bioavail 24:248–252
Shahid M, Dumat C, Silvestre J, Pinelli E (2012c) Effect of fulvic acids on lead-induced oxidative stress to metal sensitive *Vicia faba* L. plant. Biol Fertil Soils 48:689–697
Shahid M, Pinelli E, Dumat C (2012d) Review of Pb availability and toxicity to plants in relation with metal speciation; role of synthetic and natural organic ligands. J Hazard Mater 219–220:1–12
Shahid M, Ferrand E, Schreck E, Dumat C (2013a) Behavior and impact of zirconium in the soil-plant system: plant uptake and phytotoxicity. Rev Environ Contam Toxicol 221:107–127
Shahid M, Xiong T, Castrec-Rouelle M, Leveque T, Dumat C (2013b) Water extraction kinetics of metals, arsenic and dissolved organic carbon from industrial contaminated poplar leaves. J Environ Sci 25:2451–2459
Shahid M, Xiong T, Masood N, Leveque T, Quenea K, Austruy A, Foucault Y, Dumat C (2014a) Influence of plant species and phosphorus amendments on metal speciation and bioavailability in a smelter impacted soil: a case study of food-chain contamination. J Soils Sediments 14:655–665
Shahid M, Austruy A, Echevarria G, Arshad M, Sanaullah M, Aslam M, Nadeem M, Nasim W, Dumat C (2014b) EDTA-enhanced phytoremediation of heavy metals: a review. Soil Sediment Contam 23:389–416
Shahid M, Dumat C, Pourrut B, Sabir M, Pinelli E (2014c) Assessing the effect of metal speciation on lead toxicity to *Vicia faba* pigment contents. J Geochem Explor 144:290–297
Shahid M, Dumat C, Pourrut B, Silvestre J, Laplanche C, Pinelli E (2014d) Influence of EDTA and citric acid on lead-induced oxidative stress to *Vicia faba* roots. J Soils Sediments 14:835–843
Shahid M, Pinelli E, Pourrut B, Dumat C (2014e) Effect of organic ligands on lead-induced oxidative damage and enhanced antioxidant defense in the leaves of *Vicia faba* plants. J Geochem Explor. doi:10.1016/j.gexplo.2014.01.008
Shahid M, Pourrut B, Dumat C, Nadeem M, Aslam M, Pinelli E (2014f) Heavy-metal-induced reactive oxygen species: phytotoxicity and physicochemical changes in plants. Rev Environ Contam Toxicol 232:1–44
Shahid M, Dumat C, Pourrut B, Abbas G, Shahid N, Pinelli E (2015a) Role of metal speciation in lead-induced oxidative stress to *Vicia faba* roots. Russian J Plant Physiol 62:448–454
Shahid M, Khalid S, Abbas G, Shahid N, Nadeem M, Sabir M, Aslam M, Dumat C (2015a) Heavy metal stress and crop productivity. In: Hakeem KR (ed) Crop Prod Glob Environ Issues, Springer International Publishing, pp 1–25
Sheng X-F, Xia J-J (2006) Improvement of rape (*Brassica napus*) plant growth and cadmium uptake by cadmium-resistant bacteria. Chemosphere 64:1036–1042
Shi GR, Cai QS, Liu QQ, Wu L (2009) Salicylic acid-mediated alleviation of cadmium toxicity in hemp plants in relation to cadmium uptake, photosynthesis, and antioxidant enzymes. Acta Physiol Plant 31:969–977
Shi H, Ye H, Chan Z (2014) Nitric oxide-activated hydrogen sulfide is essential for cadmium stress response in bermudagrass (*Cynodon dactylon* (L). Pers.). Plant Physiol Biochem 74:99–107
Shi GL, Zhu S, Bai SN, Xia Y, Lou LQ, Cai QS (2015) The transportation and accumulation of arsenic, cadmium, and phosphorus in 12 wheat cultivars and their relationships with each other. J Hazard Mater 299:94–102

Silber A, Bar-Yosef B, Suryano S, Levkovitch I (2012) Zinc adsorption by perlite: effects of pH, ionic strength, temperature, and pre-use as growth substrate. Geoderma 170:159–167
Simmons RW, Pongsakul P, Saiyasitpanich D, Klinphoklap S (2005) Elevated levels of cadmium and zinc in paddy soils and elevated levels of cadmium in rice grain downstream of a zinc mineralized area in Thailand: implications for public health. Environ Geochem Health 27:501–511
Singh HP, Batish DR, Kaur G, Arora K, Kohli RK (2008a) Nitric oxide (as sodium nitroprusside) supplementation ameliorates Cd toxicity in hydroponically grown wheat roots. Environ Exp Bot 63:158–167
Singh S, Khan AN, Nazar R, Anjum NA (2008b) Photosynthetic traits and activities of antioxidant enzymes in blackgram (*Vigna mungo* L. Hepper) under cadmium stress. Am J Plant Physiol 3:1–25
Singh A, Sharma RK, Agrawal M, Marshall FM (2010) Health risk assessment of heavy metals via dietary intake of foodstuffs from the wastewater irrigated site of a dry tropical area of India. Food Chem Toxicol 48:611–619
Sinha S, Mukherjee SK (2008) Cadmium-induced siderophore production by a high Cd-resistant bacterial strain relieved Cd toxicity in plants through root colonization. Curr Microbiol 56:55–60
Son K-H, Kim D-Y, Koo N, Kim K-R, Kim J-G, Owens G (2012) Detoxification through phytochelatin synthesis in Oenothera odorata exposed to Cd solutions. Environ Exp Bot 75:9–15
Souza VL, de Almeida A-AF, Lima SGC, de M Cascardo JC, da C Silva D, Mangabeira PAO, Gomes FP (2011) Morphophysiological responses and programmed cell death induced by cadmium in *Genipa americana* L. (Rubiaceae). Biometals 24:59–71
Staelens J, Houle D, Schrijver A, Neirynck J, Verheyen K (2008) Calculating dry deposition and canopy exchange with the canopy budget model: review of assumptions and application to two deciduous forests. Water Air Soil Pollut 191:149–169
Stobart AK, Griffiths WT, Ameen-Bukhari I, Sherwood RP (1985) The effect of $Cd^{2+}$ on the biosynthesis of chlorophyll in leaves of barley. Physiol Plant 63(3):293–298
Stone K, Ksebati MB, Marnett LJ (1990) Investigation of the adducts formed by reaction of malondialdehyde with adenosine. Chem Res Toxicol 3:33–38
Sun Q, Ye ZH, Wang XR, Wong MH (2007) Cadmium hyperaccumulation leads to an increase of glutathione rather than phytochelatins in the cadmium hyperaccumulator Sedum alfredii. J Plant Physiol 164:1489–1498
Sun J, Shen Z (2007) Effects of Cd stress on photosynthetic characteristics and nutrient uptake of cabbages with different Cd-tolerance. Ying Yong Sheng Tai Xue Bao 18:2605–2610
Sun Y, Zhou Q, Diao C (2008) Effects of cadmium and arsenic on growth and metal accumulation of Cd-hyperaccumulator *Solanum nigrum* L. Bioresour Technol 99:1103–1110
Sun Y, Zhou Q, Wang L, Liu W (2009) Cadmium tolerance and accumulation characteristics of *Bidens pilosa* L. as a potential Cd-hyperaccumulator. J Hazard Mater 161:808–814
Tamás L, Mistrík I, Alemayehu A, Zelinová V, Bočová B, Huttová J (2015) Salicylic acid alleviates cadmium-induced stress responses through the inhibition of Cd-induced auxin-mediated reactive oxygen species production in barley root tips. J Plant Physiol 173:1–8
Tang X, Pang Y, Ji P, Gao P, Hung T (2016) Cadmium uptake in above-ground parts of lettuce (*Lactuca sativa* L.). Ecotoxicol Environ Saf 125:102–106
Tao S, Sun L, Ma C, Li L, Li G, Hao L (2013) Reducing basal salicylic acid enhances Arabidopsis tolerance to lead or cadmium. Plant and Soil 372:309–318
Tauber C (1988) Spurenelemente in Flugaschen (in German). Verlag TUV Rheinland GmbH, Köln, Germany, 469 pp
Tavarez M, Macri A, Sankaran RP (2015) Cadmium and zinc partitioning and accumulation during grain filling in two near isogenic lines of durum wheat. Plant Physiol Biochem 97:461–469

Thomine S, Wang R, Ward JM, Crawford NM, Schroeder JI (2000) Cadmium and iron transport by members of a plant metal transporter family in Arabidopsis with homology to Nramp genes. Proc Natl Acad Sci U S A 97:4991–4996

Thornton I (1992) Cadmium in the human environment. IARC, Lyon, p 169

Tian S, Lu L, Zhang J, Wang K, Brown P, He Z, Liang J, Yang X (2011) Calcium protects roots of *Sedum alfredii* H. against cadmium-induced oxidative stress. Chemosphere 84:63–69

Tipping E, Lofts S, Lawlor AJ (1998) Modelling the chemical speciation of trace metals in the surface waters of the Humber system. Sci Total Environ 210–211:63–77

Tiwari S, Kumari B, Singh SN (2008) Evaluation of metal mobility/immobility in fly ash induced by bacterial strains isolated from the rhizospheric zone of *Typha latifolia* growing on fly ash dumps. Bioresour Technol 99:1305–1310

Toppi L, Vurro E, De Benedictis M, Falasca G, Zanella L, Musetti R, Lenucci MS, Dalessandro G, Altamura MM (2012) A bifasic response to cadmium stress in carrot: early acclimatory mechanisms give way to root collapse further to prolonged metal exposure. Plant Physiol Biochem 58:269–279

Traina SJ (1999) The environmental chemistry of cadmium. In: McLaughlin MJ, Singh BR (eds) Cadmium in soils and plants. Springer, Netherlands, pp 11–37

Tremblay S, Douki T, Cadet J, Wagner JR (1999) 2-Deoxycytidine glycols, a missing link in the free radical-mediated oxidation of DNA. J Biol Chem 274:20833–20838

Tudoreanu L, Phillips CJC (2004) Modeling cadmium uptake and accumulation in plants. Adv Agron 84:121–157

Turrión MB, Lafuente F, Mulas R, López O, Ruipérez C, Pando V (2012) Effects on soil organic matter mineralization and microbiological properties of applying compost to burned and unburned soils. J Environ Manage 95 Suppl: S245–S249

Udovic M, McBride MB (2012) Influence of compost addition on lead and arsenic bioavailability in reclaimed orchard soil assessed using *Porcellio scaber* bioaccumulation test. J Hazard Mater 205–206:144–149

Ueno D, Yamaji N, Kono I, Huang CF, Ando T, Yano M, Ma JF (2010) Gene limiting cadmium accumulation in rice. Proc Natl Acad Sci U S A 107:16500–16505

United Nations Environment Programme (UNEP 2010) Final review of scientific information on cadmium. December:1–118

Unyayar S, Celik A, Cekiç FO, Gözel A (2006) Cadmium-induced genotoxicity, cytotoxicity and lipid peroxidation in *Allium sativum* and *Vicia faba*. Mutagenesis 21:77–81

Urano J, Nakagawa T, Maki Y, Masumura T, Tanaka K, Murata N, Ushimaru T (2000) Molecular cloning and characterization of a rice dehydroascorbate reductase. FEBS Lett 466:107–111

USGS (United States Geological Survey) (2015) http://minerals.usgs.gov/minerals/pubs/commodity/cadmium/. Accessed December 2015

Uzu G, Sobanska S, Sarret G, Muñoz M, Dumat C (2010) Foliar lead uptake by lettuce exposed to atmospheric fallouts. Environ Sci Technol 44:1036–1042

Vaculík M, Konlechner C, Langer I, Adlassnig W, Puschenreiter M, Lux A, Hauser M-T (2012) Root anatomy and element distribution vary between two *Salix caprea* isolates with different Cd accumulation capacities. Environ Pollut 163:117–126

Valls M, Atrian S, de Lorenzo V, Fernández LA (2000) Engineering a mouse metallothionein on the cell surface of Ralstonia eutropha CH34 for immobilization of heavy metals in soil. Nat Biotechnol 18:661–665

Van Belleghem F, Cuypers A, Semane B, Smeets K, Vangronsveld J, d'Haen J, Valcke R (2007) Subcellular localization of cadmium in roots and leaves of *Arabidopsis thaliana*. New Phytol 173:495–508

Van der Vliet L, Peterson C, Hale B (2007) Cd accumulation in roots and shoots of durum wheat: the roles of transpiration rate and apoplastic bypass. J Exp Bot 58:2939–2947

Vanhoudt N, Cuypers A, Horemans N et al (2011) Unraveling uranium induced oxidative stress related responses in *Arabidopsis thaliana* seedlings. Part II: responses in the leaves and general conclusions. J Environ Radioact 102:638–645

Verbruggen N, Hermans C, Schat H (2009) Mechanisms to cope with arsenic or cadmium excess in plants. Curr Opin Plant Biol 12:364–372

Vogel-Mikus K, Arcon I, Kodre A (2010) Complexation of cadmium in seeds and vegetative tissues of the cadmium hyperaccumulator *Thlaspi praecox* as studied by X-ray absorption spectroscopy. Plant and Soil 331:439–451

Vollenweider P, Cosio C, Günthardt-Goerg MS, Keller C (2006) Localization and effects of cadmium in leaves of a cadmium-tolerant willow (*Salix viminalis* L.): part II Microlocalization and cellular effects of cadmium. Environ Exp Bot 58:25–40

Vurro E, Ruotolo R, Ottonello S, Elviri L, Maffini M, Falasca G, Zanella L, Altamura MM, Sanità di Toppi L (2011) Phytochelatins govern zinc/copper homeostasis and cadmium detoxification in *Cuscuta campestris* parasitizing *Daucus carota*. Environ Exp Bot 72:26–33

Wang H-C, Wu J-S, Chia J-C, Yang C-C, Wu Y-J, Juang R-H (2009) Phytochelatin synthase is regulated by protein phosphorylation at a threonine residue near its catalytic site. J Agric Food Chem 57:7348–7355

Wang M, Chen L, Chen S, Ma Y (2012) Alleviation of cadmium-induced root growth inhibition in crop seedlings by nanoparticles. Ecotoxicol Environ Saf 79:48–54

Willekens H, Chamnongpol S, Davey M, Schraudner M, Langebartels C, Van Montagu M, Inze D, Van Camp W (1997) Catalase is a sink for $H_2O_2$ and is indispensable for stress defence in C3 plants. EMBO J 16:4806–4816

Williams LE, Pittman JK, Hall JL (2000) Emerging mechanisms for heavy metal transport in plants. Biochim Biophys Acta 1465:104–126

Wojas S, Hennig J, Plaza S, Geisler M, Siemianowski O, Skłodowska A, Ruszczyńska A, Bulska E, Antosiewicz DM (2009) Ectopic expression of Arabidopsis ABC transporter MRP7 modifies cadmium root-to-shoot transport and accumulation. Environ Pollut 157:2781–2789

Wójcik M, Vangronsveld J, D'Haen J, Tukiendorf A (2005) Cadmium tolerance in *Thlaspi caerulescens*: II. Localization of cadmium in *Thlaspi caerulescens*. Environ Exp Bot 53:163–171

Wong CKE, Cobbett CS (2009) HMA P-type ATPases are the major mechanism for root-to-shoot Cd translocation in *Arabidopsis thaliana*. New Phytol 181:71–78

Wu F-B, Dong J, Qian QQ, Zhang G-P (2005) Subcellular distribution and chemical form of Cd and Cd-Zn interaction in different barley genotypes. Chemosphere 60:1437–1446

Wu C, Liao B, Wang S-L, Zhang J, Li J-T (2010) Pb and Zn accumulation in a Cd-hyperaccumulator (*Viola baoshanensis*). Int J Phytoremediation 12:574–585

Wu Y, Zhou S, Chen D, Zhao R, Li H, Lin Y (2011) Transformation of metals speciation in a combined landfill leachate treatment. Sci Total Environ 409:1613–1620

Wu Z, Zhao X, Sun X, Tan Q, Tang Y, Nie Z, Hu C (2015) Xylem transport and gene expression play decisive roles in cadmium accumulation in shoots of two oilseed rape cultivars (*Brassica napus*). Chemosphere 119:1217–1223

Xian X, Shokohifard GI (1989) Effect of pH on chemical forms and plant availability of cadmium, zinc, and lead in polluted soils. Water Air Soil Pollut 45(3-4):265–273

Xiang C, Werner BL, Christensen EM, Oliver DJ (2001) The biological functions of glutathione revisited in arabidopsis transgenic plants with altered glutathione levels. Plant Physiol 126:564–574

Xiao X, Luo S, Zeng G, Wei W, Wan Y, Chen L, Guo H, Cao Z, Yang L, Chen J et al (2010) Biosorption of cadmium by endophytic fungus (EF) *Microsphaeropsis sp.* LSE10 isolated from cadmium hyperaccumulator *Solanum nigrum* L. Bioresour Technol 101:1668–1674

Xie Y, Luo H, Hu L, Sun X, Lou Y, Fu J (2014) Classification of genetic variation for cadmium tolerance in Bermudagrass [*Cynodon dactylon* (L.) Pers.] using physiological traits and molecular markers. Ecotoxicology 23:1030–1043

Xin J, Huang B, Yang Z, Yuan J, Zhang Y (2013) Comparison of cadmium subcellular distribution in different organs of two water spinach (*Ipomoea aquatica* Forsk.) cultivars. Plant and Soil 372:431–444

Xiong J, He Z, Liu D, Mahmood Q, Yang X (2008) The role of bacteria in the heavy metals removal and growth of *Sedum alfredii* Hance in an aqueous medium. Chemosphere 70:489–494

Xiong T, Leveque T, Shahid M, Foucault Y, Mombo S, Dumat C (2014) Lead and cadmium phytoavailability and human bioaccessibility for vegetables exposed to soil or atmospheric pollution by process ultrafine particles. J Environ Qual 43:1593-1600

Xiong T, Leveque T, Shahid M, Foucault Y, Mombo S, Dumat C (2014a) Lead and cadmium phytoavailability and human bioaccessibility for vegetables exposed to soil or atmospheric pollution by process ultrafine particles. J Environ Qual. doi:10.2134/jeq2013.11.0469

Xiong T-T, Leveque T, Austruy A, Goix S, Schreck E, Dappe V, Sobanska S, Foucault Y, Dumat C (2014b) Foliar uptake and metal(loid) bioaccessibility in vegetables exposed to particulate matter. Environ Geochem Health. doi:10.1007/s10653-014-9607-6

Xu Q, Min H, Cai S, Fu Y, Sha S, Xie K, Du K (2012) Subcellular distribution and toxicity of cadmium in *Potamogeton crispus* L. Chemosphere 89:114–120

Xue D, Jiang H, Deng X, Zhang X, Wang H, Xu X, Hu J, Zeng D, Guo L, Qian Q (2014) Comparative proteomic analysis provides new insights into cadmium accumulation in rice grain under cadmium stress. J Hazard Mater 280:269–278

Yamaguchi N, Mori S, Baba K, Kaburagi-Yada S, Arao T, Kitajima N, Hokura A, Terada Y (2011) Cadmium distribution in the root tissues of solanaceous plants with contrasting root-to-shoot Cd translocation efficiencies. Environ Exp Bot 71:198–206

Yılmaz DD, Parlak KU (2011) Changes in proline accumulation and antioxidative enzyme activities in *Groenlandia densa* under cadmium stress. Ecol Indicat 11:417–423

Yu H, Wang J, Fan W, Yuan J, Yang Z (2006) Cadmium accumulation in different rice cultivars and screening for pollution-safe cultivars of rice. Sci Total Environ 370(2):302–309

Zangi R, Filella M (2012) Transport routes of metalloids into and out of the cell: a review of the current knowledge. Chem Biol Interact 197:47–57

Zawoznik MS, Groppa MD, Tomaro ML, Benavides MP (2007) Endogenous salicylic acid potentiates cadmium-induced oxidative stress in *Arabidopsis thaliana*. Plant Sci 173:190–197

Zeng K, Hwang H, Yuzuri H (2002) Effect of dissolved humic substances on the photochemical degradation rate of 1-Aminopyrene and Atrazine. Int J Mol Sci 3:1048–1057

Zhang Z-C, Chen B-X, Qiu B-S (2010) Phytochelatin synthesis plays a similar role in shoots of the cadmium hyperaccumulator *Sedum alfredii* as in non-resistant plants. Plant Cell Environ 33:1248–1255

Zhang M, Liu X, Yuan L, Wu K, Duan J, Wang X, Yang L (2012) Transcriptional profiling in cadmium-treated rice seedling roots using suppressive subtractive hybridization. Plant Physiol Biochem 50:79–86

Zhang W-L, Du Y, Zhai M-M, Shang Q (2014) Cadmium exposure and its health effects: a 19-year follow-up study of a polluted area in China. Sci Total Environ 470–471:224–228

Zhao FJ, Jiang RF, Dunham SJ, McGrath SP (2006) Cadmium uptake, translocation and tolerance in the hyperaccumulator *Arabidopsis halleri*. New Phytol 172:646–654

Zhao F-J, Ma Y, Zhu Y-G, Tang Z, McGrath SP (2015) Soil contamination in China: current status and mitigation strategies. Environ Sci Technol 49:750–759

Zhou QX, Song YF (2004) Principles and methods of contaminated soil remediation. Science Press, Beijing, China

Zhou H, Zeng M, Zhou X, Liao BH, Peng PQ, Hu M, Zhu W, Wu YJ, Zou ZJ (2015) Heavy metal translocation and accumulation in iron plaques and plant tissues for 32 hybrid rice (*Oryza sativa* L.) cultivars. Plant and Soil 386:317–329

Zhu YL, Pilon-Smits EAH, Jouanin L, Terry N (1999) Overexpression of glutathione synthetase in Indian mustard enhances cadmium accumulation and tolerance. Plant Physiol 119:73–80

Zhu Z-J, Sun G-W, Fang X-Z, Qian Q-Q, Yang X-E (2004) Genotypic differences in effects of cadmium exposure on plant growth and contents of cadmium and elements in 14 cultivars of bai cai. J Environ Sci Health B 39:675–687

# Radionuclides: Accumulation and Transport in Plants

**D.K. Gupta, S. Chatterjee, S. Datta, A.V. Voronina, and C. Walther**

## Contents



Highlights

1. Bioavailability of radionuclides in environment.
2. Green plants for restoring balance/their application for radionuclide accumulation.

D.K. Gupta (✉) • C. Walther
Gottfried Wilhelm Leibniz Universität Hannover, Institut für Radioökologie und Strahlenschutz (IRS), Herrenhäuser Str. 2, Gebäude, 4113, 30419 Hannover, Germany
e-mail: guptadk1971@gmail.com; walther@irs.uni-hannover.de

S. Chatterjee • S. Datta
Defence Research Laboratory, DRDO, Post Bag 2, Tezpur 784001, Assam, India
e-mail: drlsoumya@gmail.com; sndatta1978@gmail.com

A.V. Voronina
Department of Radiochemistry and Applied Ecology, Physical Technology Institute, Ural Federal University, Mira str., 19, Ekaterinburg, Russia
e-mail: av.voronina@mail.ru



P. de Voogt (ed.), *Reviews of Environmental Contamination and Toxicology, Volume 241*, Reviews of Environmental Contamination and Toxicology 241,
DOI 10.1007/398_2016_7

3. Rhizosphere, rhizobacteria and metal transporters used in radionuclide remediation.
4. Radionuclide exposure and plant responses.
5. Radio-phytoremediation—an approach to decontaminate radionuclides using plants.

# 1 Introduction

With the discovery of X-rays in 1895 and radioactivity in the subsequent year, awareness of radiation and radioactivity became obvious and a new branch of science was started and subsequently developed into its deliberate use. However, the term radionuclide is much newer. Referring to an atomic species having precise constitution of its nucleus containing a certain number of nucleons (neutrons (N) and protons (Z) and its nuclear energy state), in 1947, Dr. T.P. Kohman of University of Chicago proposed the name nuclide. Thus, a nuclide consists of strongly bound protons and neutrons often referred to as nucleons (Kohman 1947). Nuclides may be stable and exist for an indefinite period of time (common material) or unstable. Usually, elements having atomic number higher than 83 (heavier than bismuth $^{83}Bi$), have only unstable nuclei with excess energy available. This surplus energy within a nucleus is either transformed internally or passed on to a newly formed radiation particle within the nucleus. This kind of nuclides with potential to undergo radioactive decays is called radionuclide. The decay goes together with the release of ionizing radiations like gamma ray(s) and/or subatomic, high speed alpha or beta particles. The nuclides are ultimately converted to stable nuclides. Ionizing radiation is capable of generating ions by displacing electrons in the living matter (like in DNA) and thereby disturbing its function (http://water.epa.gov/drink/contaminants/basicinformation/radionuclides.cfm, accessed 24.08.2015). However, exposure to radioactivity is a common and natural phenomenon. Apart from about 20 % anthropogenic sources (especially diagnostic imaging like X-rays, CT scans etc.) (Brenner and Hall 2007), most of the radiation exposures are usually natural (like cosmic radiation, radon gas from rocks and soil, or $^{40}K$ through foods). Water from natural sources may contain radionuclides due to dissolution of such materials from the earth crust or sporadic release from laboratories or nuclear power plants. Though, in nature, unstable nuclides with long half-lives, so called primordial radionuclides, are also present. Especially $^{235}U$ has been widely used for different applications in defence and civilian sectors. It is best used in nuclear power plants (due to low cost consumption), where, theoretically 1 kg of U (approx. 1500 tonnes of coal equivalent) can produce about 20 terajoules ($2 \times 10^{13}$ J) of energy.

However, extensive use and indiscriminate or improper disposal of wastes, insufficient decontamination at mining sites, release of production tails or decommissioning sites of such material are a serious environmental concern and can lead to contamination by radioactive substances. And, radiation is produced by spontaneous decay of radioactive materials, the amount of which and power for penetration within the body may vary with each type. However, all kinds of ionizing radiations produce health effects that may vary in different tissues and even in different individuals. Exposure to radiation and radionuclides (radiation emitters) are well-known to affect the entire body while inhalation or ingestion may affect diverse tissues inside the body (http://www.medindia.net/patients/patientinfo/radiation-hazards.htm). Ionizing radiation is capable of ionizing many atoms or molecules and destroying molecular bonds. For example, due to a nuclear explosion or accident at a nuclear power plant, radioactive caesium or iodine can be released into the environment, which may directly or indirectly get accumulated within the body, for example through the food chain. For quantification of the radionuclide uptake efficiency the soil-to-plant transfer factor (TF) is widely used to analyse radiological human dose via the ingestion pathway (Chakraborty et al. 2013). The TF is basically the concentration of element in plant (Bq $kg^{-1}$) divided by concentration of the same element in soil (Bq $kg^{-1}$). The TF may vary depending upon the element, soil pH and its texture, solid/liquid distribution coefficient, exchangeable $K^+$, organic matter and availability of the element at the root zone (Chakraborty et al. 2013). Starting from a given contamination level the total effective dose equivalent (TEDE) to individuals living in the neighbouring areas of waste-disposal/decommissioning sites can be calculated using models of biosphere and living habits. The Nuclear Regulatory Commission (NRC) has developed some sophisticated methodologies for the assessment of probable dose to man from evaluations of groundwater and soil contamination, irrigation processes, subsequent to food-chain pathways, of crops and forage and lifestyle of potentially exposed individuals (USNRC 2003; Napier et al. 2005, 2007). Above said chains, we want to focus following on soil to plant transfer of such radionuclides at contaminated sites (Fig. 1). Uptake of the radionuclides by plants depends upon several factors including mode of interaction with the materials and physiological characteristics of the species and factors like concentrations, bioavailability, and mobility of radionuclides in surface and subsurface geologic systems (Napier et al. 2005; Gupta and Walther 2014; Walther and Gupta 2015).

## 2 Natural Occurrence and Exposures

All types of soil and rocks contain uranium, the concentration of which is ranging from 0.003 mg $kg^{-1}$ in meteorites to 120 mg $kg^{-1}$ in phosphate rock. As a normal constituent of earth's crust, rock phosphate (U resources in phosphate rocks are estimated at $9 \times 10^6$ MT) deposits usually contain several million tons of uranium (U), radium (Ra), and thorium (Th) which leads to potential exposure of the environment to radiation (Gupta et al. 2014). Commonly studied U decay series involves the following steps:

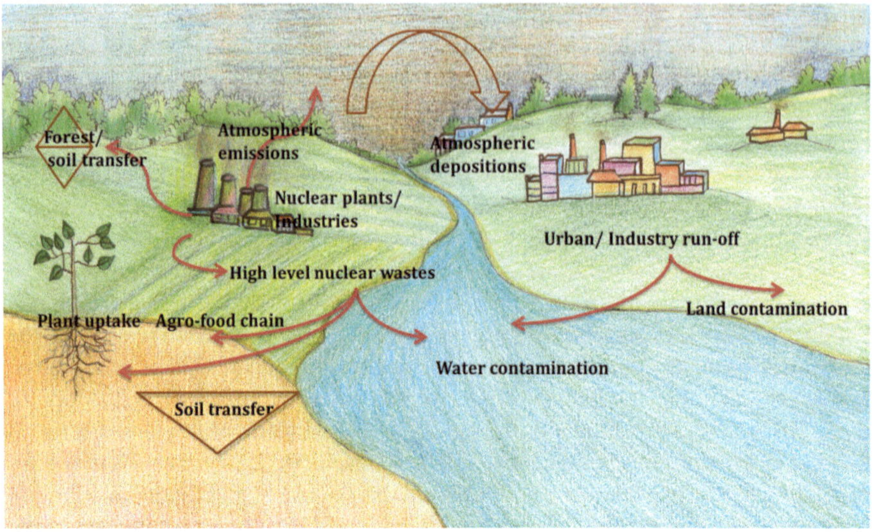


**Fig. 1** Potential radionuclide emissions and contamination pathways in the environment. Incorporation of radioactive materials may occur either as NORM (Naturally Occurring Radioactive Materials) or TENORM (Technologically Enhanced Naturally Occurring Radioactive Materials). Radioactive materials once enter into the environment gets absorbed into the soil or adsorbed to the different soil organic matter or mixed up with the water sources. Atmospheric emissions from different nuclear power plants, industries lead to deposition and pollution of the environment. Plants usually take up such elements leading to incorporation in the food chain. However, transfer factor of the radionuclides depend upon several factors including soil quality, texture and type, organic/inorganic materials, availability of elements etc

$$^{238}U \xrightarrow[\alpha]{} {}^{234}Th \xrightarrow[\beta]{} {}^{234m}Pa \xrightarrow[\beta]{} {}^{234}U \xrightarrow[\alpha]{} {}^{230}Th \xrightarrow[\alpha]{} {}^{226}Ra \xrightarrow[\alpha]{} {}^{222}Rn \xrightarrow[\alpha]{} {}^{218}Po \xrightarrow[\alpha]{}$$

$$^{214}Pb \xrightarrow[\beta]{} {}^{214}Bi \xrightarrow[\beta]{} {}^{214}Po \xrightarrow[\alpha]{} {}^{210}Pb \xrightarrow[\beta]{} {}^{210}Bi \xrightarrow[\beta]{} {}^{210}Po \xrightarrow[\alpha]{} {}^{206}Pb$$

High levels of dose rate are found above soils predominated by granite rocks or mineral sand in comparison to the other soil types. Inhabitants of communities settled on this type of soil (like monazite based sand beaches of Espirito Santo and Rio de Janeiro in Brazil) receive doses many times the average global radiation level (Eisenbud and Gesell 1997). Additional sources of radiation are up-concentrated when earth crust products (oil, coal, coal ash, minerals, and phosphate based fertilizers etc.) are extracted refined or used, which is called naturally-occurring radioactive materials (NORM). Anthropogenic activities like sewage sludge treatment, mining of radioactive materials, concentrate or expose radioactive materials that occur naturally in ores, soils, water, or other natural materials are the sources of production of NORM (IAEA 2014) or technologically enhanced NORM. Radionuclides of concern found in NORM mainly include isotopes of U, Th, Ra, Rn, Pb and Po. Reports suggest that, based on samples collected from 15 different countries, the average concentrations of $^{40}K$, $^{238}U$ and $^{232}Th$, in coal are 50, 20, and 20 Bq $kg^{-1}$ (1.35, 0.54, and 0.54 $nCikg^{-1}$), respectively (however, the concentration may vary considerably in different coal mines). Similarly, crude oil and natural gases are another source of radiation, where, approximately (average estimation), 40 $pCiL^{-1}$ (1.5 Bq $L^{-1}$) of Ra may be found at well heads of natural gas. Further, LPG (liquefied petroleum gas) processing and blending tends to enhance concentrations of radon and its products $^{210}Pb$, $^{210}Po$ (Eisenbud and Gesell 1997). Uranium contamination in soil and water has been increased in a number of regions throughout the world due to technological progressions related to establishment of nuclear power plants and its fuel cycle (e.g., mining and milling of U ore and its waste), combustion of fossil fuels (e.g., coal), increased production of phosphate fertilizers through phosphate rock mining and other uses for warheads and present as in the form of TENORM, which leads to potential ecological risks to the biota through radiation as well as chemical toxicity (Sheppard 2001).

Phosphogypsum (mostly calcium sulfate) is a by-product of a wet-acid based process during the generation of phosphoric acid in the phosphate based fertilizer industry. Large quantities of deposits of phosphogypsum are a potential source of enhanced natural radiation and for heavy metals. Approximately 70 % of calcium phosphate along with different impurities like radionuclides is present in high grade ores of phosphate rock (PR) for the production of phosphoric acid and those impurities ultimately end up in the phosphogypsum. USEPA (http://www.epa.gov/radiation/neshaps/subpartr/about.html) reports suggest the concentration of $^{238}U$ and $^{226}Ra$ in phosphogypsum of central Florida were about 10 times for U and 60 times for $^{226}Ra$ higher than the average background levels in soil. The concentration may vary according to the geographical area and quality of the PR ore. Mineral sands (sands with a specific gravity of more than 2.9), produced through the erosion of rocks (naturally or anthropogenically though blasting etc.), are also a major source of $^{232}Th$ and $^{238}U$ (UNSCEAR 1993: http://www.unscear.org/unscear/en/publications/1993.html).

# 3 Use of Green Plants for Restoring Balance and Its Applications

Phytoremediation (Ancient Greek: phyto- 'plant,' and Latin remedium- 'restoring balance'), according to a recent definition by Landmeyer (2011), is the "application of plant-controlled interactions with groundwater and organic and inorganic molecules at contaminated sites to achieve site-specific remedial goals". Cleaning up of the environment through plants is rendered by diverse environmental pollution problems through direct uptake of toxic chemical(s), followed by subsequent transformation, transport, and their accumulation in less toxic forms (Schnoor et al. 1995). Furthermore, remediation processes are being improved by plants, roots release of exudates and enzymes that induce microbial diversity at rhizosphere and biochemical activity in the bulk soil and mineralization (Macek et al. 2000; Chatterjee et al. 2013b).

Phytoremediation techniques are becoming a popular alternative to the conventional energy and instrument intensive, chemical-based expensive restoration techniques of vast polluted areas of land and water (Padmavathiamma and Li 2007; Lone et al. 2008). Use of green living systems for environmental remediation was first reported in 1948 for accumulation of nickel by the plant *Alyssum bertolonii*; however, the concept received momentum later. Since the last two decades phytoremediation work has got much attention throughout the globe (Gupta 2013).

After the reactor accident at Chernobyl, Ukraine in 1986 that caused the release of large quantities of radioactive particles into the environment and resulted in some 30 deaths mostly due to deterministic radiation effects and more due to long-term (stochastic) effects like thyroid cancers and leukaemia in the case of clean-up workers (UNSER report on 'The Chernobyl accident' from- http://www.unscear.org/unscear/en/chernobyl.html), the scientific community of the world started thinking to use plants for radionuclide decontamination of soil and water. In 1998, Consolidated Growers and Processors (CGP), PHYTOTECH, and the Ukraine's Institute of Bast Crops, employed the phytoremediation process using industrial hemp (Cannabis) as most efficient plant useful for eliminating toxins such as metals, solvents, pesticides, explosives etc. from the contaminated topsoil. Continuing to this application of plant based green technology, US Department of Defense and USEPA jointly developed plant-based clean-up approaches to large-scale clean-up projects, where, specific plants were used to remove toxins and certain metals (Rai and Pal 1999).

Use of plants for removal of inorganic components or toxic elements like Hg, Cd, As, Cr, Cs, Pb and Sr involves extraction and translocation of toxic cations or oxyanions to above ground tissues by converting the element to a less toxic chemical species (Meagher 2000; Chatterjee et al. 2012). However, for organic compounds like polychlorinated biphenyl (PCBs), polycyclic aromatic hydrocarbons (PAH), dioxin, and trichloroethylene, the objective of phytoremediation processes is to totally mineralize them into relatively nontoxic constituents, such as carbon dioxides, nitrate and ammonia (Cunningham et al. 1996). Plants deal with such components through the strategies of stabilization, exclusion, detoxification and/or their storage in specific cells or cell organelles (vacuoles, cell walls). These strategies are variable with regard to the component characteristics and the processes termed as: phytostabilization, phytoextraction, phytovolatilization, rhizofiltration, phytodegradation, and phytostimulation (Marques et al. 2009; Chatterjee et al. 2013b). Furthermore, plants also develop defence mechanisms through the production of different anti-stress proteins like metallothioneins and phytochelatins etc.

Hyperaccumulator plants, that actively uptake exceedingly large amounts of one or more heavy metals (at concentrations 100–1000-fold higher than those found in non-hyperaccumulating species) from the soil, without exhibiting any symptoms of phytotoxicity, are especially suited candidates for phytoremediation (Reeves 2006). Plants belonging to the families of Brassicaceae, Papilionaceae, Caryophyllaceae,

Poaceae and Asteraceae are most important and offer best potential for heavy metal phytoremediation. Amongst these, species belonging to the Asteraceae family show bio-removal potential of heavy metals and radionuclides, such as Sr, Cs and U. Plants like *Lactuca sativa* L., *Silybum marianum* Gaertn., *Centaurea cyanus* L., *Carthamus tinctorius* L. from Asteraceae, can uptake $^{134}Cs$ very efficiently; high biomass producing varieties of *Helianthus annuus* and *H. tuberosus* are also suitable for the phytoremediation of polluted sites (Tang and Willey 2003). Phytoremediation of radionuclides has many advantages over the traditional treatments. Firstly, in phytoremediation the soil is treated in situ, which does not cause further disruption to the soil dynamics. Secondly, plants are established, they remain for consecutive harvests to continually remove the contaminants. Last but not least, phytoremediation reduces the time of workers who are supposed to expose to radionuclides. Finally, phytoremediation can be used as a long term treatment that can provide an affordable way to restore radionuclide contaminated areas (Gupta and Walther 2014). Table 1 summarizes the successful used radionuclide phytoremediator plant species (aquatic/terrestrial).

## 4 Rhizosphere, Rhizobacteria and Metal Transporters Used in Remediation Purposes

Plant roots secrete exudates in the adjacent soil matrix at the rhizosphere (region of soil at the root-soil interface, which is subjected to root secretions and related soil microorganisms). This mechanism helps in metal-chelation to control the entry of metal within plant through the root cell. Plants root exudates normally increase the abundance of soil microflora (bacterial and fungal communities) by 1-4 orders of magnitude compared to the surrounding bulk soil that helps to degrade diverse pollutants (Anderson et al. 1994). Root exudates include: diffusates (e.g. amino or organic acids, water, inorganic ions, and sugars etc.), excretions (e.g. bicarbonates, protons, and carbon dioxide etc.), secretions (e.g. mucilage, siderophores, and allelopathic compounds etc.) which help to modulate soil microflora community composition to have a better range of metabolic capabilities with unique gene pool (Hall 2002; LeDuc and Terry 2005). The availability of U and $^{137}Cs$ to plants can be enhanced using citric acid and ammonium nitrate, respectively, in soil. However, soil augmentation may be done in a properly managed fashion, as for the inherent risks associated with the application which may in turn contaminate soil or groundwater (Prasad 2011). Plant cell walls operate as cation exchanger, sharing or excluding diverse metals, to retain the homeostasis in the cell (Rauser 1999).

On the other hand, plant growth-promoting rhizobacteria (PGPR) play a crucial role in plants growth by countering physiological stress in contaminated soils. PGPR generally performs rhizospheric colonization by stimulating the production of plant growth regulators like indole acetic acid, gibberellic acid, ethylene and cytokinins contributing to plant health and better response. Further, rhizobacteria

**Table 1** Selected plant list which are successful used as a Cs, U and Ra phytoremediator species. (After Eapen et al. 2007; Chakraborty et al. 2013; Favas and Pratas 2013; Fukuda et al. 2014; Hu et al. 2014; Favas et al. 2014)

| Plant species | Family | Radionuclide | Habitat |
|---|---|---|---|
| *Amaranthus retroflexus* | Amaranthaceae | $^{137}Cs$ | Terrestrial |
| *Apium nodiflorum* | Apiaceae | U | Aquatic |
| *Atriplex canescens* | Amaranthaceae | $^{226}Ra$ | T |
| *Batrachospermum virgato-decaisneanum* | Batrachospermaceae | $^{137}Cs$ | A |
| *Bertholletia excelsa* | Lecythidaceae | $^{226}Ra$ | T |
| *Brassica juncea* | Brassicaceae | $^{137}Cs$ | T |
| *Brassica oleracea* | Brassicaceae | $^{137}Cs$ | T |
| *Callitriche stagnalis* | Plantaginaceae | U | A |
| *Carex buxbaumii* | Cyperaceae | $^{226}Ra$ | T |
| *Carthamus tinctorius* | Asteraceae | $^{134}Cs$ | T |
| *Centaure acyanus* | Asteraceae | $^{134}Cs$ | T |
| *Chloroidium saccharophilum* | Chlorophyta | $^{137}Cs$ | T |
| *Dryopteris scottii* | Dryopteridaceae | $^{226}Ra$ | T |
| *Eleocharis dulcis* | Cyperaceae | U | A |
| *Fontinalis antipyretica* | Fontinalaceae | U | A |
| *Helianthus annuus* | Asteraceae | U | T |
| *Helianthus tuberosus* | Asteraceae | U | T |
| *Hydrilla verticillata* | Hydrocharitaceae | U | A |
| *Iris pseudacorus* | Iridaceae | $^{226}Ra$ | T |
| *Juncus inflexus* | Juncaceae | $^{226}Ra$ | T |
| *Lactuca sativa* | Asteraceae | $^{134}Cs$ | T |
| *Lemna aoukikusa* | Tracheophyta | $^{137}Cs$,U | A |
| *Lemna gibba* | Araceae | U | A |
| *Lemna minor* | Araceae | U | A |
| *Nymphaea violacea* | Nymphaeaceae | $^{137}Cs$ | A |
| *Phalaris arundinacea* | Poaceae | $^{137}Cs$ | T |
| *Phaseolus acutifolius* | Fabaceae | U | T |
| *Phragmites australis* | Poaceae | $^{228Th}$, $^{226}Ra$ | T |
| *Pteridium aquilinum* | Pteridaceae | $^{226}Ra$ | T |
| *Pteris multifida* | Pteridaceae | $^{226}Ra$ | T |
| *Silybum marianum* | Asteraceae | $^{134}Cs$ | T |
| *Typha latifolia* | Typhaceae | U, $^{226}Ra$ | A |
| *Zostera japonica* | Zosteraceae | U | A |
| *Zostera marina* | Zosteraceae | U | A |

may also secrete antibiotics, phosphate solubilising substances, hydrocyanic acid, siderophores, 1-aminocyclopropane-1-carboxylic acid (ACC) to increase bioavailability and root absorption of different metals (Meyer 2000; Davies et al. 2001). In a nickel-contaminated soil, Burd et al. (1998) reported that after the addition of *Kluyvera ascorbata* SUD165/26 an associated rhizobacteria, germination and growth of Indian mustard (*Brassica juncea*) seeds increase by 50–100 % with respect to the control plants.

Root cell plasma membranes harbour a number of metal transporter-proteins which play important roles in heavy metal homeostasis. Among different families of transporters, NRAMP (natural resistance-associated macrophage protein), ZIP (Zinc importer) families (ZRT, IRT-like Protein; [ZRT-Zinc regulated transporter, IRT-iron regulated transporter]), CDF (cation diffusion facilitator) family, heavy metal ATPases (HMAs) family like P1B-ATPases, copper transporter (COPT) family proteins, ATP-binding cassette (ABC) transporters, ABC transporters of the mitochondria (ATM), $Ca^{2+}$ cation antiporter (CAX), multidrug resistance-associated proteins (MRP), yellow-stripe-like (YSL) pleiotropic drug resistance (PDR) transporters are well studied transporters (Dubey 2011; Huang et al. 2012; Gupta et al. 2013). Histidine-rich domain of ZIP family transporters is supposed to get activated in response to divalent metals and their uptake. As for example, AtZIP4 (*Arabidopsis thaliana* ZIP4) proteins help in Zn transport and Cd uptake from soil to root cells and Cd transport from root to shoot (Krämer et al. 2007). While, transport of Fe, and other metals like $Mn^{2+}$, $Zn^{2+}$, and $Cd^{2+}$ in root cells was being carried out by IRT1 in *A. thaliana* (Nishida et al. 2008). HMAs family transporters (P1B-type ATPases) are basically internal transporters that load Cd and Zn metals from the surrounding tissues into the xylem and perform as an efflux pump (Krämer et al. 2007). AtHMA3 transporter of tonoplast membrane in *A. thaliana* sequesters a wide variety of heavy metals and its over-expression raises the tolerance to heavy metals like Cd, Co, Pb, and Zn (Manara 2012; Gupta et al. 2013). Strategies for plants to isolate metals from dynamic cytosol and metabolically active cellular compartments include vacuolar sequestration using proton pumps like vacuolar proton-ATPase (V-ATPase) and vacuolar proton-pyrophosphatase (V-PPase) (Dalcorso et al. 2010).

## 5 Radio-Phytoremediation: An Approach to Decontaminate Radionuclides Using Plants

Radionuclides and/or toxic elements contaminated soils, sediments, surface water and groundwater's remediation by using green plants is widely reported (Nishita et al. 1958; IAEA 1989; Soudek et al. 2004, 2006). However, radionuclide decontamination has its inherent problem of radioactivity itself. Plants require a long period of time to contact with a contaminant to evolve the ability to hyperaccumulate radioactive material (Fig. 2). The concentrations, mobility, and bioavailability of radionuclides depend upon several factors. These include the quality, quantity and the rate of release of radionuclides present at the source; hydrological factors, like dispersion, advection, and dilution; geochemical processes, such as complexation at aqueous phase, pH, adsorption/desorption, solid/liquid distribution coefficient, reduction/oxidation (redox), ion exchange,

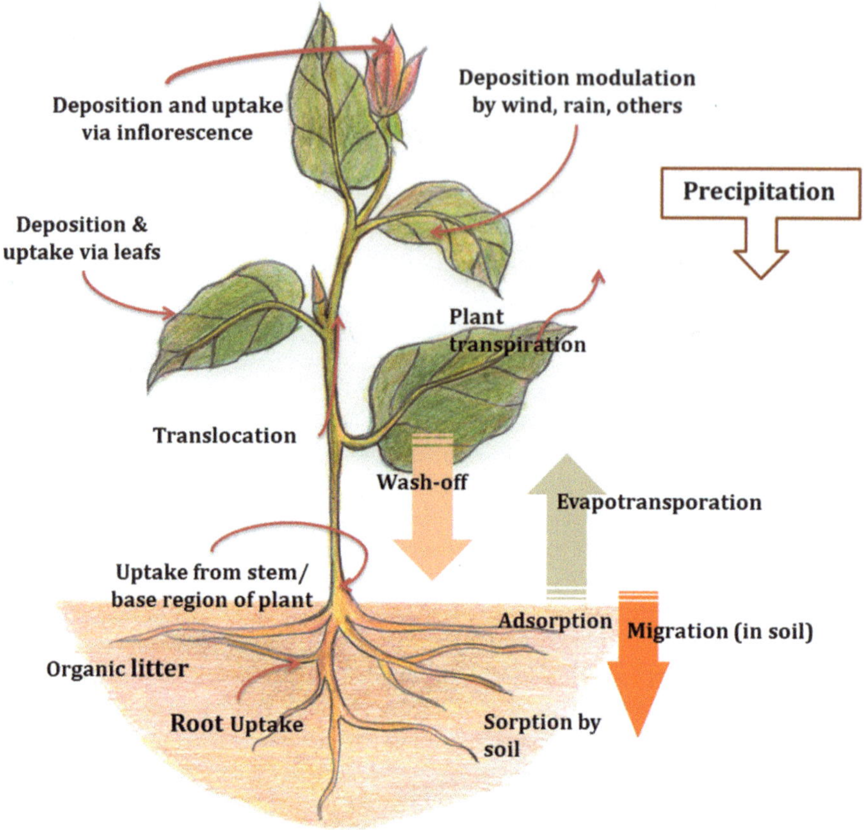


**Fig. 2** Radionuclide contamination pathways in plants. The root uptake of radionuclide depends upon various factors like the condition of metal with clay/soil particle/minerals and its availability to the plants. The translocation and deposition of the same may take place in the leaf or inflorescence. Again, metals present in the atmosphere also get deposited on different plant body surfaces (like leaf surfaces). The concentration of deposition varies and modulated through the factors like wind, rain etc. Wash-off metals further deposited into soil and sorbed by soil particles, and available for plants uptake

precipitation/dissolution, diffusion, colloid-facilitated transport, exchangeable potassium ion distribution, anion exclusion and organic matter contents (Albrecht et al. 2002; Napier et al. 2012; Chakraborty et al. 2013; Hegazy et al. 2013). Absorption and distribution of contamination in plants may take place either through direct (exposures at aerial organs like leaf, stem, and tendrils) or indirect (through root systems in soil related contamination) routes, which varies considerably in different plant species especially in case of long-lived radionuclides (Din et al. 2010). Further, soil properties/texture (like drying and subsequent cracking of soils) due to biological activity, colloid-facilitated transport in soil may augment the mobility and/or affectivity of radionuclides (Napier et al. 2012).

In the environment, a number of radionuclides (e.g. Tc, U and Pu) may be present in more than one oxidation state, and adsorption and precipitation behaviour of individual states may vary considerably. Thus, transfer pathways and efficiencies of radionuclides vary widely according to (1) chemical nature and reactivity of the isotope that may affect the solubility and transport of the isotope through soil pore (includes pore structures) water within the root (rhizosphere) of the plant, (2) exposure route (e.g. root versus foliar exposure) of the contaminant, (3) plant itself (species, physical stature, age, root structure, and root-shoot ratio etc.), (4) requirements and availability of nutrient(s) of the plant (chemical resemblance of the isotope to a nutrient) and (5) level of toxins and pathogens in the soil.

It is depending upon the water cycle and the compositions of the flora which are intrinsically coupled, thus interactions may occur within these water bodies or in terrestrial ecosystems. However, the distribution and productivity of terrestrial vegetation is mostly dependent upon the equilibrium of water. Plant communities and root zone (rhizosphere) of the plant along with the soil microorganisms play important roles in evapo-transpiration generation of runoffs (Gerten et al. 2004), metal uptake and remediation. Ehlken and Kirchner (2002) suggested that the concentration of a radionuclide in a plant or plant part is linearly related to its concentration in soil within the rhizosphere (same kind of relation is also true for transfer of heavy metals and pesticides in plants) (Gast et al. 1988; Trapp et al. 1990). However, interestingly, soil-to-plant transfer factors for a number of long-lived radionuclides (as for example, radiocesium uptake from agro-soil) may vary considerably up to three orders of magnitude (Frissel 1992; Nisbet and Woodman 2000; Ehlken and Kirchner 2002). Again significant variation is also present in case of longevity of plant roots in different plant species, their development and function is being responded by genetically determined range to their environment (Clausnitzer and Hopmans 1994).

## 6 Bioavailability of Radionuclides

In soil, most of the elements are usually present, from where plants acquire essential elements as micronutrients. Bioavailability of trace substances in soil is a complex phenomenon, where competitions of varied substances along with metal-metal interactions take place. Transfer dynamics from soil to plant is usually calculated for trace substances whose behaviour in the soil-plant-rhizosphere system largely depends on the concentrations of macro-nutrients present. It is reported that an activity concentration in a soil solution of 1 $BqL^{-1}$ of $^{90}Sr$ or $^{137}Cs$ corresponds to ca. $2 \times 10^{-15}$ $MolesL^{-1}$ though the average concentrations of chemical homologes K, Ca, and Mg, respectively, in soil solution are in the order of 1 mMoles $L^{-1}$ (Robson and Pitman 1983; Ehlken and Kirchner 2002). Uptake of radionuclides like radioactive caesium and strontium and heavy metals by plants is thus affected by several factors naturally present in the soil (Lorenz et al. 1994). Cation exchange reactions in the soil milieu determine the actual concentration of an ion in soil-solution which is competitive in nature. Accordingly, co-precipitation also depends on the concentrations of competing substances in solution (as for example, sorption of radiostrontium is dominated by reversible exchange with major cations like $Ca^{2+}$). While strontium is exchanged in preference to Ca in minerals, the preference of the same metal reverses in the presence of organic matter (Valcke and Cremers 1994). The sorption of radiocaesium in soils is determined by ion exchange to a few sites, as for example, weathered mica which are accessible only to poorly hydrated cations and shows high selectivity for $Cs^+$ over $K^+$ and $NH_4^+$ (Comans and Hockley 1992; Ehlken and Kirchner 2002). In a recent report by Voronina et al. (2015), they reported a comparative study of selectivity and reversibility of radiocaesium and radiostrontium sorption by natural aluminosilicates (glauconite and clinoptilolite) as well as modified ferrocyanide sorbents, it was shown from their experiment that modification of natural aluminosilicates by ferrocyanides allows to increase distribution coefficients of caesium by 100–1000 times, to improve sorption capacity and to make sorption of caesium more selective and almost irreversible. They recommended the use of modified aluminosilicates for remediation of radioactively contaminated lands.

Some metals (like Zn, Cu) are essential for plant growth and sustenance, however, higher levels of essential or non-essential metals cause toxicity and inhibition of growth in most of the plants (Hall 2002; Lasat 2002). Absorption in plants is a vital issue, where the root zone (rhizosphere) region along with soil microorganisms interacts with the elements for bio-availability and uptake of the same (Wenzel et al. 2003). Elements are typically co-transported across the plasma membrane in the form of cations. Toxic heavy metals which do not have any known biological function (viz. As, Cd, and Pb) are also transported through the common transportation system (Manara 2012). Generally- in plants, root secrete exudates in adjacent soil matrix which may help in chelation of unwanted metals and to prevent transportation of these metals inside the cell (Marschner 1995). As for example, root exudates like histidine (His) prevent Ni uptake from the soil (Salt et al. 2000), whereas, pectic sites and different extracellular carbohydrate molecules present on the cell wall play an important role in immobilization of toxic heavy metal ions (Manara 2012). In metal contaminated soil, plants act upon as either "accumulators" or "excluders" for particular metal(s), where the excluders restrict metal uptake into their biomass and accumulators amass (high concentration) metals or contaminants into their aerial tissues like leaf tissues after breakdown of the same (Tangahu et al. 2011). However, chemical forms of metals may vary as bound component to soil particle and/or precipitated form, which mostly make a major portion of the metal(s) that is insoluble or unavailable to plants (Chatterjee et al. 2013a). Several factors involve in controlling uptake, translocation, and storage of different metals including toxic elements into the plant body. In soil microorganisms, plant induce pH changes and redox reactions, plants also produce chelating agents and exudates and a group of transporters embedded in plant cell plasma membrane, it is the major regulator for several functions. The metal transporter proteins like proton pumps –ATPases (that generate electrochemical gradients and consume energy), co-and anti-transporters proteins (that use the electrochemical gradients generated by –ATPases), and ion channel proteins (that facilitate the transport of ions into the cell) involved in metal uptake and translocation (Tangahu et al. 2011).

## 7 Radionuclide Exposure and Plant Responses

Plants have been observed to take up many cations present in their root region irrespective of biological requirement. When plants are exposed to ionizing radiation, molecular and cellular effects are induced directly through damage of macromolecules or indirectly through water radiolytic reactions producing reactive oxygen species (ROS) (Gupta and Walther 2014). By energy transfer from the radiation field to plant tissue, ionizing radiation can directly induce DNA strand breaks, lipid oxidation, or enzyme denaturation. ROS are also produced under natural metabolism and their functions in plants are as signaling molecules, regulating normal growth and development and they are useful in stress responses. Plants also possess an antioxidative defense system comprising enzymes (e.g., superoxide dismutase (SOD) and catalase (CAT)) and metabolites (e.g., ascorbate and glutathione) to regulate the amount of ROS in cells (Gupta and Sandalio 2012). Plant metal tolerance mechanisms requires the coordination of complex physiological and biochemical processes. Plants employ various strategies to cope with toxic effects of radionuclides and heavy metals or metalloids, in general, and radionuclides stress either by avoidance (restricting the metal uptake), or by tolerance (survive in the presence of high internal metal concentration). Plants also restrict metal stress by the mechanisms like reducing the concentration of metal entering into the cell by extracellular precipitation, biosorption to cell walls, reduced uptake, and/or by increased efflux (Gupta and Walther 2014). However, tolerating metal stress involves elaborate physiological responses like intracellular chelation through synthesis of amino acids, organic acids, glutathione (GSH), and/or by metal-binding ligands such as metallothioneins (MTs) and phytochelatins (PCs), vacuolar compartmentalization, and up-regulation of antioxidant defense and glyoxalase systems to counter the deleterious effects caused by ROS (Gupta and Sandalio 2012).

## 8 Phytoremediation of Selected Radionuclides

### *8.1 Caesium*

Caesium (Cs) has the chemical similarity to potassium (K) and thus absorption and translocation by plants is common. Contamination of Cs ($^{134}$Cs half-life:$T_{1/2} = 2.06$ years and $^{137}$Cs -$T_{1/2} = 30.04$ years) leads to long persistence in soil and environment (Koarashi et al. 2012; Kamei-Ishikawaa et al. 2013). This alkaline metal is present in solution as free hydrated cation $Cs^+$. However, plant uptake of Cs was first reported by Collander in early 1940s (Collander 1941) and was thought to be coupled as a nutrient analogue with the same $K^+$ uptake transporter, especially during low K concentrations (Menzel 1954; Zhu and Smolders 2000). It is apparent that $K^+$ is the most significant cation amongst all other alkaline metals and compounds that competes with $Cs^+$ uptake (Zhu and Shaw 2000). Plant families such as Brassicaceae, Amaranthaceae and Chenopodiaceae are the major Cs accumulators. For plant uptake of Cs, $K^+$ concentration in soil is the most important factor. Thus, both selection of plant and potassium concentration in the soil should be taken into consideration for phytoremediation of Cs from soil (Buysse et al. 1996).

Cs uptake in plants is not yet been fully elucidated on a microscopic level. However, it has been observed that, low K concentration in soil helps to uptake Cs through K uptake system. Thus, higher K concentration suppresses the Cs uptake (Zhu and Smolders 2000). Studies on discrimination factor on Cs/K revealed that plant roots usually absorb Cs less ably than K like nutrient analogue (Smolders et al. 1996). Two $K^+$ transport pathways ($K^+$ channels and $K^+$ transporters) are involved with $Cs^+$ transportation, where multiple genes may be involved. As for example, in *A. thaliana*, high affinity $K^+$ transporter family (e.g. AtHAK5) and voltage-insensitive cyclic nucleotide gated channels, AtCNGCs, are involved in the $Cs^+$ uptake (Kanter et al. 2010; Kobayashi et al. 2010; Yamashita et al. 2014).

Specific absorption mechanism in soil by plant may depend upon several factors like availability and exchangeability of Cs along with the particular salt content and its interionic effects, soil type, water submergence time etc. (Tensho et al. 1961, Mimura et al. 2001). Salts like ammonium molybdophosphate (AMP) and rubidium chloride augment in uptake of Cs from soil. The application of AMP for partitioning and recovery of radioactive $^{137}Cs$ from $HNO_3$ and $NaNO_3$ rich liquid waste is well documented (Mimura et al. 2001).

However, a number of conditions and physiological parameters of plants are involved that lead to differential uptake in different species. Among these factors, plants demand for potassium and its rooting pattern for its growth strategies are most important, along with the factors like presence of mycorrhizal fungi, rate of root growth, efficiency of ion transporters that are present on plasma membranes of root cells etc. (Broadley and Willey 1997; Zhu and Smolders 2000). Halophytes of family Chenopodiaceae are long been known as Cs accumulators, however, their capability to discriminate K and Na from Cs for uptake from salty soil is still discussed controversially (Flowers et al. 1986; Zhu and Smolders 2000).

To understand the genetic variations in plants for uptake of Cs is very important for selection of suitable crops grown in soils and their accumulation pattern within the edible parts. This kind of data may be helpful to minimize its transfer to food chain. Further, non-edible plants may also be explored to decontaminate the soil contaminated with this radionuclide (Zhu and Smolders 2000). Reports showed that red root pigweed (*Amaranthus retroflexus*) is a good accumulator of $^{137}Cs$ with higher uptake rate and plant growth, as also shown during the phytoremediation practices for Cs-contaminated soil in the vicinity of Chernobyl, Ukraine

(Lasat et al. 1997, 1998; Dushenkov et al. 1999). Kang et al. (2012) showed that napier grass (*Pennisetum purpureum* Schum.) under hydroponic condition is also a good accumulator of Cs and may be applied for phytoremediation. To reduce the total time required for phytoremediation, various chemical and biological amendments may be useful in speeding up phytoextraction process. Reports suggest that the application of $NH_4NO_3$, $(NH_4)_2SO_4$ could increase phytoextraction, while application of potassium based components to the soil should be minimized (Lasat et al. 1997, 1998; Dushenkov et al. 1999; Zhu and Smolders 2000). Further, genetic alteration and special plant breeding by selection of suitable plant taxa may be helpful to achieve the goal.

In a recent report by Yamashita et al. (2014) on the deposition of radionuclides (radiocesium- $^{134}$Cs, $^{137}$Cs) in different plant species after the accident of the Fukushima Daiichi Nuclear Power Station, Japan in March 2011, it is shown that the transfer factors of radiocesium from soil to plant([Cs]plant/[Cs]soil) vary widely in different plants. However, plants like *Athyriumyo koscense*, *Dryopteris tokyoensis* and *Cyperus brevifolius* exhibited relatively high TF values (more than 0.4).

## 8.2 *Uranium*

U has six naturally known isotopes $^{233}$U to $^{238}$U with half-lives varying between 69 years and 4.47 billion years. Among three primordial natural isotopes, $^{238}$U is the most abundant isotope (~99.275 % of the uranium found in nature) followed by $^{235}$U (~0.720 %), and $^{234}$U (~0.005 %). Radioactivity and metal toxicity of U makes it a pollutant of concern to the environment. Naturally, U is released into water or soil through the weathering of rocks in the oxidized zone of the terrestrial near-surface environment. In an almost constant isotopic ratio $^{238}$U(137.88):$^{235}$U(1), release of U to water takes place through natural geochemical process, while $^{234}$U is produced by the radioactive decay of $^{238}$U. A range of physicochemical forms like free metal ion ($U^{4+}$ or $UO_2^{2+}$), inorganic compounds (e.g., uranyl carbonate or uranyl phosphate), mineralogical matrices and soil humic substances (e.g., uranyl fulvate or humate) in dissolved, colloidal, and/or particulate forms regulate the speciation and bioavailability of U (Markich 2002). Biologically dissimilatory metal reducing bacteria (DMRB) and sulfate reducing bacteria (SRB) play an important role in U(VI) reduction to U(IV) in anaerobic environments (Liu et al. 2002; Payne et al. 2002). Though U has no known essential biological functions, it is taken up by a variety of mineral and other surfaces by the presence of dissolved calcium by inducing the formation of ternary uranyl-calcium carbonate complexes. However, redox state of U distribution is a consequence of the redox potential in solution, predominantly the U(VI) to U(IV) ratio, usually regulate its solubility and potential for migration and formation of soluble $UO_2$ (Fredrickson et al. 2000; Stewart 2008). Organic acids like citric acid were found to be most effective in increasing U accumulation in plants. Huang et al. (1998) showed that U accumulation in shoot of *Brassica juncea* and *B. chinensis* increased from less than 5 mg $kg^{-1}$ to more than

5000 mg $kg^{-1}$ in citric acid-treated soils, where soil concentration of total U was 750 mg $kg^{-1}$.

Chemical and civil engineering based U remediation have several apprehensions for its high economic and energy implications, chemicals that are used for the process and their fates, specificity on the characteristics of contaminant, etc. On the other hand, phytoextraction of uranium contaminated soil and water have recently gained importance as it provides a low-cost alternative to currently employed remediation procedures. Phytoremediation processes of uranium generate a minimum amount of secondary waste(s) that may otherwise produce large amounts of heavy metal laden leachate. Results of the study by Huang and his colleagues for phytoextraction of uranium through amendment of soil using citric acid, however, showed the process to be highly effective in triggering U hyperaccumulation in selected plant species (Huang et al. 1998). The application of citric acid transiently reduced the soil pH that enhances desorption of soil U leading to more root absorption. Favas and Pratas (2013) reported that plants like *Callitriche stagnalis*, *Apium nodiflorum* and *Fontinalis antipyretica* have the capacity to accumulate significant amounts of U. These authors also suggested *C. stagnalis*, *A. nodiflorum*as are keystone species which can be used for different phytoremediation applications for removal of U from the contaminated sites for their efficient rooting capability, high biomass and bio-productivity and the positive correlation with the uranium concentration in waters (Favas and Pratas 2013; Pratas et al. 2012). Duckweed (*Lemna gibba* and *L. minor*) was also reported to have good ability to accumulate U in the standing water sources (Mkandawire and Dudel 2005; Favas and Pratas 2013). Further, greenhouse experiments with *B. juncea* and *H. annuus* showed the phytoremediation of uranium-contaminated soil is enhanced in *B. juncea* when the soil is augmented with citric acid buffer (pH 4.8) that caused an enhancement of the desorption of U in the soil (Dushenkov 2003; Huhle et al. 2008).

## 8.3 Radium

Radium ($^{226}$Ra), discovered by Marie Curie and Pierre Curie in 1898, is the product of $^{238}$U decay series but is 2.7 million times more radioactive than the same molar amount of natural U due to its shorter half-life. In the last nine steps of the fourteen steps $^{238}$U decay series, Ra decay (Ra emanation) occurs producing exradio ($^{222}$Ra gas), Ra A ($^{218}$Po), Ra B ($^{214}$Pb), Ra C ($^{214}$Bi) etc. until stable $^{206}$Pb is reached. Ra is a very important radionuclide due to its (inclusive isotopes $^{226}$Ra and $^{228}$Ra) presence in all three natural decay series, relatively long half-lives. It has a high mobility in different environmental conditions like uranium mining and processing industries, phosphate and gold mining areas. Ra is a well-known stressors of human body and accumulates in bone (IAEA 2002, 2014). In general, naturally four Ra isotopes are present in the environment: uranium decay series- $^{226}$Ra, thorium decay series $^{228}$Ra and $^{224}$Ra and actinium decay series $^{223}$Ra. Among these, $^{226}$

Ra and $^{228}$Ra have half-lives of 1600 and 5.75 years, respectively (IAEA 2002, 2014). Alkaline earth metal Ra is a readily reactive element that reacts with soil and its organic matter. However, absorption of Ra by plants from root zone decreases with high soil sulphate content, increasing pH and increased exchangeable Ca (IAEA 2014). $^{226}$Ra is mostly studied among its other isotopes. In plants, grown in natural environment, Ra concentrations vary from 0.0044 to 52 Bq $kg^{-1}$ (DW), with higher concentrations reported in Brazil nut (Simon and Ibrahim 1990).

Wetland plants like *Typha latifolia*, *Phragmites australis*, *Juncus inflexus*, *Carex buxbaumii* and, *Iris pseudacorus* showed interesting result in the accumulation of Ra from a constructed wetland contaminated with U/Ra heavy metals (Soudek et al. 2006, 2007). Macrophytic algae, members of Characeae family have also been shown as hyper-accumulators of radium, the mechanism of which needs to be ascertained properly (Kalin et al. 2002). However, Ra may be loaded into the calcium-rich lattice and/or form of $BaSO_4$ substitute crystals in algal tissue (Kalin et al. 2002; Kunze et al. 2007).

## 9 Summary and Conclusion

The two most crucial factors for successful implementation of specific plants for restoration of radionuclide contaminated sites are (1) ability to uptake the radioactive material to relatively high levels (2) without affecting the growth or high biomass production. Potential plants should usually accumulate radionuclide in the above ground parts at a level that exceed the soil concentration. There are several factors that limit the application of plant based techniques for remediation, like the rapid incorporation of the metal within/in clay/soil particle/minerals which may limits its availability to the plants. However, phytoremediation might be the most valuable easy technique for restoration of radionuclide contaminated soil if selected cultivars are used. Further, selection of plants, genetic manipulation of different transporters, elucidation of anti-stress factors in plant physiology are the main future area of comprehensive research that will prove useful to counter our own poised nuclear age.

**Acknowledgement** Authors are thankful to Mrs. Swagata Chatterjee for making handmade figures. S.C. and S.D. are thankful to Director, DRDO, Assam, India. The authors apologize for the many colleagues who are not referenced in this work due to space limitations.

## References

Albrecht A, Schultze U, Liedgens M, Flühler H, Frossard E (2002) Incorporating soil structure and root distribution into plant uptake models for radionuclides: toward a more physically based transfer model. J Environ Radioact 59:329–350

Anderson TA, Kruger EL, Coats JR (1994) Enhanced degradation of a mixture of three herbicides in the rhizosphere of a herbicide-tolerant plant. Chemosphere 28:1551–1557

Brenner DJ, Hall EJ (2007) Computed tomography—an increasing source of radiation exposure. New Engl J Med 357:2277–2284
Broadley MR, Willey NJ (1997) Differences in root uptake of radiocaesium by 30 plant taxa. Environ Pollut 95:311–317
Burd GI, Dixon DG, Glick BR (1998) A plant growth-promoting bacterium that decreases nickel toxicity in seedlings. Appl Environ Microbiol 64:3663–3668
Buysse J, Van de Brande K, Merckx R (1996) Genotypic differences in the uptake and distribution of radiocaesium in plants. Plant Soil 178:265–271
Chakraborty SR, Azim R, Rahman AKMR, Sarker R (2013) Radioactivity concentrations in soil and transfer factors of radionuclides from soil to grass and plants in the Chittagong city of Bangladesh. J Phys Sci 24:95–113
Chatterjee S, Datta S, Mallick PH, Mitra A, Veer V, Mukhopadhyay SK (2013a) Use of wetland plants in bioaccumulation of heavy metals. In: Gupta DK (ed) Plant based remediation processes, soil biology, vol 35. Springer, Berlin Heidelberg
Chatterjee S, Mitra A, Datta S, Veer V (2013b) Phytoremediation protocols: an overview. In: Gupta DK (ed) Plant based remediation processes, soil biology, vol 35. Springer, Berlin Heidelberg
Chatterjee S, Singh L, Chattopadhyay B, Datta S, Mukhopadhyay SK (2012) A study on the waste metal remediation using floriculture at East Calcutta Wetlands, a Ramsar site in India. Environ Monit Assess 184:5139–5150
Clausnitzer V, Hopmans JW (1994) Simultaneous modelling of transient three dimensional root growths and soil water flow. Plant Soil 164:299–314
Collander R (1941) Selective absorption of cations by higher plants. Plant Physiol 16:691–720
Comans RNJ, Hockley DE (1992) Kinetics of cesium sorption on illite. Geochim Cosmochim Acta 56:1157–1164
Cunningham SD, Anderson TA, Schwab P, Hsu FC (1996) Phytoremediation of soils contaminated with organic pollutants. Adv Agron 56:55–114
Dalcorso G, Farinati S, Furini A (2010) Regulatory networks of cadmium stress in plants. Plant Signal Behav 5:1–5
Davies FT Jr, Puryear JD, Newton RJ (2001) Mycorrhizal fungi enhance accumulation and tolerance of chromium in sunflower (*Helianthus annuus*). J Plant Physiol 158:777–786
Din KS, Harb S, Abbady A, Saad N (2010) The distribution of the radionuclides Ra-226, Th-232 and K-40 in various parts of the Alfalfa plant. Tenth radiation physics and protection conference, 27–30 November, Nasr City-Cairo, Egypt. http://www.rphysp.com/s7/distribution.pdf
Dubey RS (2011) Metal toxicity, oxidative stress and antioxidative defense system in plants. In: Gupta SD (ed) Reactive oxygen species and antioxidants in higher plants. CRC Press, Boca Raton, FL
Dushenkov S (2003) Trends in phytoremediation of radionuclides. Plant Soil 249:167–175
Dushenkov S, Mikheev A, Prokhnevsky A, Ruchko M, Sorochisky B (1999) Phytoremediation of radiocaesium-contaminated soil in the vicinity of Chernobyl, Ukraine. Environ Sci Technol 33:469–475
Eapen S, Singh S, D'Souza SF (2007) Phytoremediation of metals and radionuclides. In: Singh SN, Tripathi RD (eds) Environmental bioremediation technologies. Springer, Germany
Ehlken S, Kirchner G (2002) Environmental processes affecting plant root uptake of radioactive trace elements and variability of transfer factor data: a review. J Environ Radioact 58:97–112
Eisenbud M, Gesell T (1997) Environmental radioactivity (fourth edition) from natural, industrial and military sources. Academic, USA
Favas PJC, Pratas J (2013) Uptake of uranium by native aquatic plants: potential for bio indication and phytoremediation. E3S Web of Conferences1.13007. doi: 10.1051/e3sconf/20130113007; Article available at http://www.e3s-conferences.org or http://dx.doi.org/10.1051/e3sconf/20130113007
Favas PJC, Pratas J, Varun M, D'Souza R, Paul MS (2014) Accumulation of uranium by aquatic plants in field conditions: prospects for phytoremediation. Sci Total Environ 470–471:993–1002

Flowers TJ, Hajibagheri MA, Clipson NJW (1986) Halophytes. Quart Rev Biol 61:313–337
Fredrickson JK, Zachara JM, Kennedy DW, Duff MC, Gorby YA, Li SMW, Krupka KM (2000) Reduction of U(VI) in goethite (alpha-FeOOH) suspensions by a dissimilatory metal-reducing bacterium. Geochim Cosmochim Acta 64:3085–3098
Frissel MJ (1992) An update of the recommended soil-to-plant transfer factors of Sr-90, Cs-137 and transuranic. In: International Union of Radio Ecologists (ed) VIIIth report of the working group soil to-plant transfer factors. Balen, Belgium, pp 16–25, IUR Pub R-9212-02
Fukuda S, Iwamoto K, Atsumi M, Yokoyama A, Nakayama T, Ishida K, Inouye I, Shiraiwa Y (2014) Global searches for microalgae and aquatic plants that can eliminate radioactive cesium, iodine and strontium from the radio-polluted aquatic environment: a bioremediation strategy. J Plant Res 127:79–89
Gast CH, Jansen E, Bierling J, Haanstra L (1988) Heavy metals in mushrooms and their relationship with soil characteristics. Chemosphere 17:789–799
Gerten DS, Schaphoff S, Haberlandt U, Lucht W, Sitch S (2004) Terrestrial vegetation and water balance–hydrological evaluation of a dynamic global vegetation model. J Hydrol 286:249–270
Gupta DK (2013) Plant based remediation process. Springer, Germany. ISBN 978-3-642-35563-9
Gupta DK, Sandalio LM (2012) Metal toxicity in plants: perception, signaling and remediation. Springer, Germany. ISBN 978-3-642-22080-7
Gupta DK, Huang HG, Corpas FJ (2013) Lead tolerance in plants: strategies for phytoremediation. Environ Sci Pollut Res 20:2150–2161
Gupta DK, Chatterjee S, Datta S, Veer V, Walther C (2014) Role of phosphate fertilizers in heavy metal uptake and detoxification of toxic metals. Chemosphere 108:134–144
Gupta DK, Walther C (2014) Radionuclide contamination and remediation through plants. Springer, Germany. ISBN 978-3-319-07664-5
Hall JL (2002) Cellular mechanism for heavy metal detoxification and tolerance. J Exp Bot 53:1–11
Hegazy AK, Afifi SY, Alatar AA, Alwathnani HA, Emam MH (2013) Soil characteristics influence the radionuclide uptake of different plant species. Chem Ecol 29:255–269
Hu N, Ding D, Li G, Zheng J, Li L, Zhao W, Wang Y (2014) Vegetation composition and 226Ra uptake by native plant species at a uranium mill tailings impoundment in South China. J Environ Radioact 129:100–106
Huang J, Zhang Y, Peng JS, Zhong C, Yi HY, Ow DW, Gong JM (2012) Fission yeast HMT1 lowers seed cadmium through phytochelatin-dependent vacuolar sequestration in *Arabidopsis*. Plant Physiol 158:1779–1788
Huang JW, Blaylock MJ, Kapulnik Y, Ensley AD (1998) Phytoremediation of uranium-contaminated soils: role of organic acids in triggering uranium hyperaccumulation in plants. Environ Sci Technol 32:2004–2008
Huhle B, Heilmeier H, Merkel B (2008) Potential of *Brassica juncea* and *Helianthus annuus* in phytoremediation for uranium. In: Merkel BJ, Hasche-Berger A (eds) Mining and hydrogeology. pp 307–318
IAE (International Atomic Energy) (1989) Clean-up of large areas contaminated as a result of a nuclear accident. Technical reports series no. 300
IAEA (International Atomic Energy Agency) (2002) Non-technical factors impacting on the decision making processes in environmental remediation, IAEA-TECDOC- 1279. IAEA, Vienna
IAEA (International Atomic Energy Agency) (2014) The environmental behaviour of radium: revised edition, technical reports. Series no 476. IAEA, Vienna
Kalin M, Kießig G, Küchler A (2002) Ecological water treatment processes for underground uranium mine water: Progress after three years of operating a constructed wetland. In: Merkel BJ, Planer-Friedrich B, Wolkersdorfer C (eds) Uranium in the aquatic environment. Proceedings of the International Conference Uranium Mining and Hydrogeology III and the International Mine Water Association Symposium, Freiberg, Germany, 15-21, September 2002, Springer Verlag, Berlin, Heidelberg, New York, pp 587

Kamei-Ishikawaa N, Itob A, Tagamic K, Umitaa T (2013) Fate of radiocesium in sewage treatment process released by the nuclear accident at Fukushima. Chemosphere 93:689–694

Kang DJ, Seo YJ, Saito T, Suzuki H, Ishii Y (2012) Uptake and translocation of cesium-133 in napier grass (*Pennisetum purpureum* Schum.) under hydroponic conditions. Ecotoxicol Environ Saf 82:122–126

Kanter U, Hauser A, Michalke B, Draxl S, Schaffner AR (2010) Caesium and strontium accumulation in shoots of *Arabidopsis thaliana*: genetic and physiological aspects. J Exp Bot 61:3995–4009

Koarashi J, Moriya K, Atarashi-Andoh M, Matsunaga T, Fujita H, Nagaoka M (2012) Retention of potentially mobile radiocesium in forest surface soils affected by the Fukushima nuclear accident. Sci Rep 2:1005

Kobayashi D, Nozomi N, Hisamatsu S, Yamagami M (2010) At KUP/HAK/KT9, a K? Transporter from *Arabidopsis thaliana*, mediates Cs? Uptake in *Escherichia coli*. Biosci Biotechnol Biochem 74:203–205

Kohman TP (1947) Proposed new word: nuclide. Am J Phys 15:356–357

Krämer U, Talke IN, Hanikenne M (2007) Transition metal transport. FEBS Lett 581:2263–2272

Kunze C, Kießig G, Küchler A (2007) Management of passive biological water treatment systems for mine effluents. In: Marmiroli N, Borys S, Marmiroli M (eds) Advanced science and technology for biological decontamination of sites affected by chemical and radiological nuclear agents. Springer, Germany

Landmeyer JE (2011) Introduction to phytoremediation of contaminated groundwater. Springer, London

Lasat MM (2002) Phytoextraction of toxic metals: a review of biological mechanisms. J Environ Qual 31:109–120

Lasat MM, Ebbs SD, Kochian LV (1997) Potential for phytoextraction of $^{137}$Cs from contaminated soils. Plant Soil 195:99–106

Lasat MM, Ebbs SD, Kochian LV (1998) Phytoremediation of a radiocaesium-contaminated soil: evaluation of caesium-137 bioaccumulation in shoots of three plant species. J Environ Qual 27:165–169

LeDuc DL, Terry N (2005) Phytoremediation of toxic trace elements in soil and water. J Ind Microbiol Biotechnol 32:514–520

Liu CX, Gorby YA, Zachara JM, Fredrickson JK, Brown CF (2002) Reduction kinetics of Fe(III), Co(III), U(VI), Cr(VI), and Tc(VII) in cultures of dissimilatory metal-reducing bacteria. Biotechnol Bioengg 80:637–649

Lone MI, He ZH, Stoffella J, Yang X (2008) Phytoremediation of heavy metal polluted soils and water: progresses and perspectives. J Zhejiang Univ Sci B 9:210–220

Lorenz SE, Hamon RE, McGrath SP, Holm PE, Christensen TH (1994) Applications of fertilizer cations affect cadmium and zinc concentrations in soil solutions and uptake by plants. Eur J Soil Sci 45:159–165

Macek T, Mackova M, Kas J (2000) Exploitation of plants for the removal of organics in environmental remediation. Biotechnol Adv 18:23–34

Manara A (2012) Plant responses to heavy metal toxicity. In: Furini A (ed) Plants and heavy metals, Springer briefs in biometals. Springer, New York, pp 27–53

Markich SJ (2002) Uranium speciation and bioavailability in aquatic systems: an overview. Sci World J 2:707–729

Marques APGC, Rangel AOSS, Castro PML (2009) Remediation of heavy metal contaminated soils: phytoremediation as a potentially promising clean-up technology. Crit Rev Environ Sci Technol 39:622–654

Marschner H (1995) Mineral nutrition for higher plants, 2nd edn. Academic Press and Harcourt Brace and Co., London-San Diego-New York-Boston-Sydney-Tokyo-Toronto

Meagher RB (2000) Phytoremediation of toxic elemental and organic pollutants. Curr Opin Plant Biol 3:153–162

Menzel RG (1954) Competitive uptake by plants of potassium, rubidium, cesium, and calcium, strontium, barium from soils. Soil Sci 77:419–426
Meyer JM (2000) Pyoverdines: pigments siderophores and potential taxonomic markers of fluorescent Pseudomonas species. Arch Microbiol 174:135–142
Mimura H, Saito M, Akiba K, Onodera Y (2001) Selective uptake of cesium by ammonium molybdophosphate (AMP)-calcium alginate composites. J Nucl Sci Technol 38:872–878
Mkandawire M, Dudel EG (2005) Accumulation of arsenic in *Lemna gibba* L. (duckweed) in tailing waters of two abandoned uranium mining sites in Saxony, Germany. Sci Total Environ 336:81–89
Napier BA, Krupka KM, Valenta MM, Gilmore TJ (2005) Soil and groundwater sample characterization and agricultural practices for assessing food chain pathways in biosphere model. U.S. Nuclear Regulatory Commission, Washington, DC, NUREG/CR-6881
Napier BA, Fellows RJ, Krupka KM (2007) Soil-to-plant concentration ratios for assessing food chain pathways in biosphere models. U.S. Nuclear Regulatory Commission, Washington, DC, NUREG/CR-6941
Napier BA, Fellows RJ, Krupka KM (2012) Soil-to-plant concentration ratios for assessing food chain pathways in biosphere models. U.S. Nuclear Regulatory Commission, Washington, DC, NUREG/CR-7120
Nisbet AF, Woodman RFM (2000) Soil-to-plant transfer factors for radiocesium and radiostrontium in agricultural systems. Health Phys 78:279–288
Nishida S, Mizuno T, Obata H (2008) Involvement of histidine-rich domain of ZIP family transporter TjZNT1 in metal ion specificity. Plant Physiol Biochem 46:601–606
Nishita H, Steen AJ, Larson KH (1958) Release of strontium-90 and cesium-137 from vina loam upon prolonged cropping. Soil Sci 86:195–201
Padmavathiamma PK, Li LY (2007) Phytoremediation technology: hyper-accumulation metals in plants. Water Air Soil Pollut 184:105–126
Payne RB, Gentry DA, Rapp-Giles BJ, Casalot L, Wall JD (2002) Uranium reduction by *Desulfovibrio desulfuricans* strain G20 and a cytochrome C3 mutant. Appl Environ Microbiol 68:3129–3132
Prasad MNV (2011) A state-of-the-art report on bioremediation, its applications to contaminated sites in India. Ministry of Environment & Forests, Government of India. http://www.moef.nic.in/downloads/public-information/BioremediationBook.pdf
Pratas J, Favas PJC, Paulo C, Rodrigues N, Prasad MNV (2012) Uranium accumulation by aquatic plants from uranium-contaminated water in Central Portugal. Int J Phytorem 14:221–234
Rai UN, Pal A (1999) Toxic metals and phytoremediation. Enviro News, Newsletter of International Society of Environmental Botanists, India, vol 5(4)
Rauser WE (1999) Structure and function of metal chelators produced by plants: the case for organic acids, amino acids, phytin and metallothioneins. Cell Biochem Biophys 31:19–48
Reeves RD (2006) Hyperaccumulation of trace elements by plants. In: Morel JL, Echevarria G, Goncharova N (eds) Phytoremediation of metal-contaminated soils, NATO Science series IV: Earth and Environmental Sciences. Springer, New York
Robson AD, Pitman JB (1983) Interactions between nutrients in higher plants. In: Läuchli A, Bieleski RL (eds). Inorganic plant nutrition. New York: Springer, pp 147–180. (Encyclopedia of Plant Physiology, 1)
Salt DE, Kato N, Kräme U, Smith RD, Raskin I (2000) The role of root exudates in nickel hyperaccumulation and tolerance in accumulator and nonaccumulator species of *Thlaspi*. In: Terry N, Banuelos G (eds) Phytoremediation of contaminated soil and water. CRC Press LLC, Boca Raton
Schnoor JL, Licht LA, McCutcheon SC, Wolfe NL, Carreira LH (1995) Phytoremediation of organic and nutrient contaminants. Environ Sci Technol 29:318–323
Sheppard SC (2001) Toxicants in the environment: bringing radioecology and ecotoxicology together. In: Brechignac F, Howard BJ (eds) Radioactive pollutants: impact on the environment. EDP Sciences, Les UlisCedex A, France

Simon SL, Ibrahim SA (1990) Biological uptake of radium by terrestrial plants, vol 1, The environmental behaviour of radium. IAEA, Vienna, pp 545–599, Technical reports series no. 310

Smolders E, Kiebooms L, Buysse J, Merckx R (1996) 137Cs uptake in spring wheat (*Triticum aestivum* L. Cv. Tonic) at varying K supply. II: The effect in solution culture. Plant Soil 18:211–220

Soudek P, Tykva R, Vaková R, Vank T (2006) Accumulation of radioiodine from aqueous solution by hydroponically cultivated sunflower (*Helianthus annuus* L.). Environ Exp Bot 57:220–225

Soudek P, Petrová S, Benesová D, Tykva R, Vanková R, Vanek T (2007) Comparison of $^{226}$Ra nuclide from soil by three woody species *Betula pendula, Sambucus nigra* and *Alnus glutinosa* during the vegetation period. J Environ Radioact 97:76–82

Soudek P, Tykva R, Vank T (2004) Laboratory analyses of Cs uptake by sunflower, reed and poplar. Chemosphere 55:1081–1087

Stewart BD (2008) The dominating influence of calcium on the biogeochemical fate of uranium. Ph.D. thesis, Stanford University, USA

Tang S, Willey NJ (2003) Uptake of 134Cs by four species from the Asteraceae and two varieties from the Chenopodiaceae grown in two types of Chinese soil. Plant Soil 250:75–81

Tangahu BV, Abdullah SRS, Basri H, Idris M, Anuar N, Mukhlisin M (2011) A review on heavy metals (As, Pb, and Hg) uptake by plants through phytoremediation. Int J Chem Eng 939161

Tensho K, Yeh KL, Mitsui S (1961) The uptake of strontium and cesium by plants from soil with special reference to the unusual cesium uptake by lowland rice and its mechanism. Soil Sci Plant Nutr 6:176–183

Trapp S, Matthies M, Scheunert I, Topp EM (1990) Modelling the bioconcentration of organic chemicals in plants. Environ Sci Technol 24:1246–1252

USNRC: U.S. Nuclear Regulatory Commission, Washington, DC (2003) Literature review and assessment of plant and animal transfer factors used in performance assessment modelling. Pacific Northwest National Laboratory. NUREG/CR-6825, PNNL-14321

Valcke E, Cremers A (1994) Sorption-desorption dynamics of radiocesium in organic matter soils. Sci Total Environ 157:275–283

Voronina AV, Blinova MO, Semenischev VS, Gupta DK (2015) Returning lands, contaminated as a result of radiation accidents, to farming use. J Environ Radioact 144:103–112

Walther C, Gupta DK (2015) Chemistry of radionuclides in the environment: influence of chemical speciation and plant uptake on radionuclide migration. Springer, Germany. ISBN 978-3-319-22170-0

Wenzel WW, Unterbrunner R, Sommer P, Sacco P (2003) Chelate assisted phytoextraction using canola (*Brassica napus* L.) in outdoors pot and lysimeter experiments. Plant soil 249: 83–89

Yamashita J, Enomoto T, Yamada M, Ono T, Hanafusa T, Nagamatsu T, Shoji Sonoda S, Yamamoto Y (2014) Estimation of soil-to-plant transfer factors of radiocesium in 99 wild plant species grown in arable lands 1 year after the Fukushima 1 Nuclear Power Plant accident. J Plant Res 127:11–22

Zhu YG, Shaw G (2000) Soil contamination with radionuclides and potential remediation. Chemosphere 41:121–128

Zhu YG, Smolders E (2000) Plant uptake of radiocaesium: a review of mechanisms, regulation and application. J Exp Bot 51:1635–1645

# Index





P. de Voogt (ed.), *Reviews of Environmental Contamination and Toxicology Volume 241*, Reviews of Environmental Contamination and Toxicology 241,
DOI 10.1007/978-3-319-46945-4

MIX
Papier aus verantwortungsvollen Quellen
Paper from responsible sources
FSC® C105338

If you have any concerns about our products,
you can contact us on
**ProductSafety@springernature.com**

In case Publisher is established outside the EU,
the EU authorized representative is:
**Springer Nature Customer Service Center GmbH**
**Europaplatz 3, 69115 Heidelberg, Germany**

Printed by Libri Plureos GmbH
in Hamburg, Germany